This book is dedicated to the women and men of the British Columbia Forest Branch, which became the BC Forest Service, for their public service protecting the people of British Columbia and our forests.

We especially acknowledge the hundreds of Forest Branch and Forest Service staff and non-staff members who worked together on the ground, in the air and on the water, who lost their lives in service, with a recognition of the tragic consequences for their families, friends and colleagues. They are not forgotten.

Contents

Preface

This book began as a general discussion of the important contribution of the staff of the BC Forest Service with the BC Forest Service Centennial Committee and the general support of the membership of the BC Forest History Society. It would not have been possible without the unrestricted access provided by the staff of *Forests, Lands and Natural Resource Operations*. We would particularly like to acknowledge the fine work of the Records Management Centre and especially Penny Newton and Lorraine Skinner and also the assistance of Louis Syrette also of *Forests, Lands and Natural Resource Operations*. The virtual resources of the library staff, and their use of digital delivery, were world class.

Many current and former staff shared their insights with us in an extensive series of interviews and/or offered to assist us with manuscript feedback. In this regard we would very much like to thank Doug Adderley, Randy Aitken, Mike Apsey, Michael Blackstock, David Borth, Dave Caughlin, Jim Challenger, Ken Chantler, Diane Clearwater, Gordon Comeau, Lloyd Davies, Rod Deboise, Harry Doberstein, John Flanagan, Dana Hayden, Tracy Hendry, Glen Heyes, Jenji Konishi, Doug Konkin, Tom Lacey, Madeline Maley, Pat MacFee, Rob MacLaren, Steve Newton, John Parminter, Larry Pedersen, Bob Perchie, Peter Poland, Lyle Resh, Steve Schell, Tim Sheldan, Jim Snetsinger, Dirk Trigg and Phil Youwe.

This project would not have been possible without the generous support of the Weston family and their foundation, who supported the hundreds of hours of interviews and transcriptions, research and travel we undertook in recording this significant part of Canada's history. Their assistance to the Royal BC Museum and their concern about the impact of a changing climate on the landscape and biota of Canada and the world is deeply appreciated.

We would also like to acknowledge the assistance of librarians at the University of British Columbia, the University of Victoria, and the BC Legislative Library's Reference Services. Long hours of research were undertaken by history students Michelle Chan, Katie Impett and Patrick Craibe, and Dr Jenny Clayton researched and wrote materials for several sidebar entries. Archival surveys were also conducted by Chris Hanna and Sharon Keen.

We wish to thank the staff of the BC Archives for their superb support with records and research questions answered by librarian Fredericke Verspoor and the archives reference room staff. We owe a special debt to the team of colleagues involved in the Cold Storage project to maximize the survival lifespan of negatives, including those of the BC Forest Service, for their patience and assistance in accommodating our requests within their own pressing schedule. We also owe a particular debt to Kelly-Ann Turkington, our permissions officer, who dealt with a seemingly endless list of requests for images.

Throughout all this project we have been supported by the executive of the Royal BC Museum, including CEO Professor Jack Lohman, Dr Kathryn Bridge and Provincial Archivist Emeritus Gary Mitchell. Thank you all.

Any errors are our own responsibility. Any success is due to this group of people who believed that this story of the service of so many should be told.

Introduction

The creation of the British Columbia Forest Branch in 1912 marked the start of over a century of service by the many women and men who served as stewards of British Columbia's forests. It is a remarkable story. This book honours those decades of service and presents a small portion of the many events and personalities that mark the evolution of British Columbians' relationship with their forest history.

The creation of the BC Forest Service out of the Forest Branch began in part because large-scale forest change came late to British Columbia. When it did arrive, it arrived fully formed as industrial timber harvesting, both in terms of technology and the scale of capital. True, there were pioneering moments and earlier small-scale operations, but industrial forestry and external markets presented an immediate regulatory challenge to government and an economic opportunity for government revenues. At this same moment the profession

Canadian Western Lumber Co. Ltd, Fraser Mills; bucking after fallers, about 1910. B-08397.

of forestry in North America, the science and practice of understanding and managing forest lands, was establishing its North American identity.[1] The public was also entering into a debate about how to wisely manage forests for the future. New terms were being used, words like *conservation* and phrases like *natural resource policy*. It was a unique moment, and the Forest Branch was an inspiring solution. But what led to its creation?

A Slowly Changing World

When the Napoleonic wars (1803–1815) were creating massive demands for the timber of eastern Canada and the Maritimes, the forests of British Columbia sustained First Nations families and villages. They supplied sea and river canoes of Western Redcedar or cottonwood, planks for shelter, objects of cultural meaning large and small, clothing, and the tools needed to fish, process and store food. There was wealth, poverty, feast and famine. Trade with explorers brought them iron, which changed technologies for working with wood, but the impact of European contact on the forest was limited at first – unlike the impact upon the health and well-being of First Nations village life.

Further east, big changes were beginning, which, three-quarters of a century later, would have a profound effect upon both First Nations lands and British Columbia's forests. The great Canadian forest historian A.R.M. Lower argued that Canada's forest industry was created by Napoleon. The French naval blockade of the Baltic was intended to starve the British navy of strategic timber. They responded with a massive investment in eastern Canadian forests. That great scythe of demand cut its way in a slow-moving arc through the "Timber Colony" of New Brunswick, along the St Lawrence and into the Ottawa Valley, gathering logs for export at the timber coves of Quebec City. Fleets of specialized timber ships were built to service this market, and Red pine, White pine and other species floated down the Ottawa and St Lawrence rivers to export. Settlers came next, many of them war veterans. They continued the land-clearing by axe and horse, moving inland from Lake Ontario and creating farmland from forest, selling their timber and burning stumps for potash to sell to the fledgling chemical industry. Gradually, they cut their way across northern Ontario and down to Lake Superior. Large entrepreneurial firms emerged: John Booth, James McLaren, the House of William Price and a system of shantymen and resource railways feeding rivers and lakes with cut trees for market. By the 1890s these new firms had reached the Lake of the Woods on their way to the prairies.

In the northern United States the industry marched inland – driven by settlement, canals and then rail – across Maine to Vermont and Wisconsin, cutting across the south of the Great Lakes to Chicago and what became the

Victoria Lumber and Manufacturing Lumbering camp No. 5, west of Ladysmith, 1907. C-06449.

great cut-over lands of Minnesota. Armies of forest entrepreneurs, loggers, raftsmen, river drivers, railroaders and sawmill owners, as well as mill workers, families, farmers and merchants who supplied them, all moved quickly, year by year, across increasingly marginal agricultural land as great forests fell.

Both forest industry armies closed around the western end of the Great Lakes, from the north and south. And so the 1890s brought the great question to both sides of the border: where next? Many Americans went to work in the forests of the southern US. Others made the jump by rail to the west coast, to California's Redwoods or Oregon's forests at Coos Bay, or they cut along the railway grant forests from Potlatch, Idaho, to Seattle, Washington, and built the great mills of Puget Sound.

Canadian investors arrived by Canada's new transcontinental railway – the tracks built on wooden ties, crossing high wooden trestle bridges – to look for opportunities in Canada's newest province, British Columbia. Some brought their fortunes and invested in large mills, such as James McLaren's operations at what became Fraser Mills; others represented British or American investors, establishing the giant Victoria Lumber and Manufacturing Company mill at Chemainus, on Vancouver Island, or the Powell River pulp mill. Still others simply speculated and sought to acquire parcels they could offer to investors with operations back east. All sought timber. What they found in BC was a forest whose size exceeded the total of all the forests managed today by the entire US Forest Service.

But change had already been happening in BC's forest, on a smaller scale than would occur in the 1890s. European impacts on the forest were fairly limited before the 1860s. The first European use of BC timber was for mast repairs by Captain Cook in 1778 at Yuquot, in Nootka Sound. In 1788 John Meares took a shipment of spars for export to China from Nootka.[2] Timber was used to construct dwellings, posts and stockades by the North West Company (which arrived in 1803) and, after the fur trade merger, by the Hudson's Bay Company. As a proprietary colony, the Hudson's Bay Company encouraged a small stave trade for the export of salmon in barrels, but it failed to establish an export in naval supplies. There were small experiments, with minor success, in Pacific Rim lumber exports to Chile, Mexico, California, Hawaii and Australia,[3] but most were opportunistic cargoes and not a sustained export industry.

In the Crown colony (proclaimed in 1858) indigenous forest rights were limited to timber stands identified by First Nations as required for their use, as were villages, fields, and fishing and cemetery sites. The idea of use formed the core of the Indian Reserve Commission process for the remainder of the 19th century – a process that was at times fair and, in some hands, a process of

New Westminster waterfront between 1866 and 1870. A-01595.

economic, cultural, ecological and political dispossession, as shown by hundreds of hours of land claims testimony over the last decades. The forest that had nurtured First Nations for centuries, with Western Redcedar and fir-shaded salmon streams, began to be punctuated by firebreak clear-cuts, whether around small settlements based on fur posts of the 1840s or larger administrative centres such as those that emerged in the 1850s at Fort Victoria, Fort Langley and New Westminster.

The earliest Hudson's Bay Company sawmills, cutting for local consumption after failed export attempts, had difficulty maintaining production even when times were good, because "gold fever" began to draw off their workforce. The 1858–1862 gold rush brought intense environmental change, even destruction, along the streams and river banks that descended from the gold fields of BC's Cariboo region. The gravel beds where salmon spawned were dug, and river banks were washed apart with water monitors (a high pressure nozzle for cutting soil banks apart) into muddy waves channelled into crudely made wooden sluices. A series of roadhouses, ranches and instant wooden towns were created by small sawmills that next turned to making fence and boardwalk lumber, pit poles, mining timbers and the raw materials to shape hulls around steam engines to create paddle-wheelers. While intense in its local impact, the change had minimal impact on the larger forests.

Managing a Colonial Forest

The role of government in regulating the forests of British Columbia began slowly, because with such vast forests and few markets, trees had a low value. They were actually seen as impeding agricultural settlement. Settlers were required to clear lands as part of pre-emption. Through the Department of Lands, the forest lease became the mechanism to attract capital and to create settlement as much as employment. It was a utilitarian management, intended to assist settlement and attract capital.

The majority of the colonial forests were reserved for the Crown, but initial land sales, begun in 1859, included timber with land. The encouragement of colonial manufacturing was endorsed in the terms of section 53 of the 1865 Land Ordinance. Representing English capital, Captain Edward Stamp negotiated for timber cutting rights with the colonial government in 1862, importing a complete industrial sawmill, with mechanics and labourers, direct from England into traditional Nuu-chah-nulth territory on the Alberni Canal, Vancouver Island. William Banfield aided the mill's establishment and confided to the colonial secretary that he would not bring the local Chief to Victoria to meet Governor Douglas, as "the Chief is not worth bringing as he would try to impede settlement but the next in rank an excellent Indian has done all towards

Logging Burrard Inlet at the site of Denman Street, English Bay, now Vancouver, 1868. A-03391.

the quiet and easy arrangement."[4] And so sawmilling for export began to transform the inlet, ecologically and politically. Stamp's mill quickly exhausted easily accessible timber along the waterfront of the long inlet.[5]

One should not conclude that conflict characterized all relations between the forest industry and First Nations. Robert McDonald cites an Indian Reserve Commission report from 1877 that sawmill labour and other economic activity put between $80 and $100,000 into the economy of Burrard Inlet First Nations, both Squamish and Musqueam.[6] Gordon Hak calculated that in the 1870s, First Nations loggers supplied perhaps 30 per cent of the log supply of one of the province's largest sawmills, at Moodyville on Burrard Inlet (what is now the harbour of Vancouver).[7] And in 1891, in addition to working in the Moodyville sawmill, two-thirds of the port's stevedores were First Nations men and women, working in family crews.[8] In 1886 First Nations loggers held three-quarters of the provincial general timber hand-logging licences before the railway brought in a flood of new applicants for a growing number of licences.[9]

Stamp and others, like Sewell Prescott Moody (Moodyville Sawmills) and John Hendry (Royal City Planning Mills), established major export mills, most of them on Burrard Inlet and at Chemainus. There were also a number of minor pioneer sawmills established in proximity to European settlements or adjunct to the mining frontier.

The selling of cutting rights for spars while reserving the land for the Crown became an established principle in 1865. For settlers, removal of timber was necessary as part of meeting the conditions of a pre-emption and offered income through sales to mill owners. The regulatory aspect of this in British Columbia was uneven. Robert Cail argued that while the *Land Act* of 1884 forbade the sale of timber lands, it was not effectively enforced until 1906.[10] In 1870, just prior to BC's entry into Canada, section 47 of the 1870 Land Ordinance stated that conveyance of land included trees.

It was the 1885 arrival of the railway that changed the dynamic of sawmilling in BC. To finance the railway's construction, the province agreed to cede to the federal government and the railway a belt of land 20 miles (32 km) on each side of the line. An additional cessation of land was executed in the northeastern Peace River region. A belt of land 40 miles (64 km) wide ran through the south of the province, administered by the Dominion of Canada's own forest service. The land was returned to the province's jurisdiction in 1930, as were many federally administered lands in western Canada.

Moody's Mill, North Vancouver, about 1864. A-05604.

The sailing barque *Thermopylae* at the Brunette Sawmill, New Westminster. A-00592.

The local market for lumber expanded dramatically with the construction of Vancouver. Fires caused by settlement and logging operations swept the remaining forests of the lower mainland year after year with increasing frequency,[11] and the *Bush Fire Act* went through numerous unsuccessful revisions.

Confrontations arose between pre-emptors and timber licence holders over cutting rights. Yet on this relatively new regulatory frontier, those conflicts were resolved quickly. Some pre-emptors sought timber rights for themselves. Others opposed the renewal of a specific timber licence. One group of farmers at Seymour Creek, North Vancouver, sought not the timber but the right to conduct the actual logging on their lands.[12] Generally, these were minor conflicts, in part due to the clear resolution of timber rights between pre-emptors and licence holders that was set out in the 1888 revised *Land Act*.

Another reason for the minimal conflict was spatial. As population flowed into the Fraser Valley, the forests fell quickly. The scale of the mills, like Captain Stamp's at Alberni decades before, quickly outstripped the local timber supply. The shift of logging north to Jervis Inlet was symptomatic of the cheap, water-based transport of logs. Industry quickly moved beyond the range of established settlements and into the coastal forests – outside easy reach of the farmer. When the next great wave of immigration arrived in the early 20th century, agriculture and the fruit farming boom carried people into the interior, not the forest.

In the settled farming and fishing regions of the southern coast, farm settlement often pre-dated industrial forestry, and landowners had completed their own clearances. The combination of tidewater booming grounds and the small Shay and Climax logging locomotives allowed industry to move out behind settled areas, with minimal conflict, to access interior timber stands. The coastal railroads, being company-specific, were closed to competitive use by other companies or to settlers seeking timber.

Railways changed everything, attracting smaller operators to cut along the great 40-mile (64 km) wide swath of railway belt that surrounded the new railway. Some of them prospered, providing lumber for more railway ties and great wooden bridges and trestles, for building the wooden paddle-wheelers that supplied the mines along the narrow interior lakes, and for boards for fledgling towns that rose and fell with mining cycles or re-emerged from devastating fires with a brick permanency.

This was also a country of fire, and the province had to do something to manage it. Some fires had natural causes, as hundreds of lightning strikes burst

Five-spot, 0-4-2 saddle tank locomotive of the BC Mills Timber and Trading Company at Granite Bay, 1911. Note the spark arrester on the smoke stack. C-04571.

George Lovatt's sawmill Sandon, 1898. Hundreds of small mobile mills operated across the province. F-09965.

on dry grasslands and forests. Other fires were human set, whether to improve hunting, bring up traditional indigenous food plants, burn off lands for planting introduced crops, expand grazing or, according to some, reveal the rock beneath for prospectors. But the biggest source of human-caused fire was the instrument of greatest change to early forestry: the railway locomotive, with its trail of sparks meeting dry summer forest in those early years. The 1874 *Bush Fire Act* was the province's first legislative attempt to deal with fire, a problem which continued (and still continues) to be a significant worry for those engaged in forestry.

Two Forest Industries: Coast and Interior

Two styles of industry began to emerge, each with its own identity. On the coast, with its rain-fed forests, economies of scale rapidly gave the advantage to investments in large mill operations. At first the difficulty of logging mountainous terrain and along small, steep, rocky rivers constrained coastal forestry production. Later, technology would offer solutions. The steep-slope-geared Shay locomotive and steam donkey engine, with its network of cables, would allow loggers to strip hillsides and bring logs to water for rafting. From early beginnings in 1863, large export sawmills became established at Chemainus (1863) and along Burrard Inlet, such as the Hastings Mill (1867). Examples like the Pioneer Mill's first export to Australia in 1865 from Moodyville supplanted the scattered global markets initially pioneered by the Alberni operation. British Columbia began sending finished lumber to the housing markets of Australia, the Hawaiian Islands, Java, Chile and, for a brief time in the 1860s, Shanghai, China.

While centralization and scale quickly stripped the surrounding forests, the distinctive character of the coastal industry was also set in motion. A network of contract loggers and camps developed, as part of an ever-moving, far-flung series of operations that fed logs by water to central processing mills. As the major buyers of logs, they also set the price to the logger. And as Thomas Roach noted, that pattern would hold into the 1950s[13].

In the interior, scale played out in a different way. Here it was sawmill and small business operators who felt victims of scale. Large-scale eastern investors had taken the best of the Dominion railway belt timber berths from the 10-mile swath of alienated land on either side of the transcontinental railway grant. The Canadian Pacific Railway dominated as the sole market for railway ties and as the only shipper for lumber sent to market. Networks of smaller, regional operators met regularly to set prices, discuss freight costs and protest timber costs that were unfairly high, in their opinion, because their trees were of a smaller diameter and less valued species in the marketplace than the Douglas-

Saddle tank locomotive bringing logs to water, Hastings Company at Granite Bay, 1911. C-04540.

Logging in the East Kootenays, stacked for shipping, 1914. C-05912.

firs of the coastal rainforests. Vulnerable to economic downturns, the interior depended on a handful of local markets: railway buyers, settlers and the boom-and-bust (sometimes ghost town) activities related to local mining. But by 1900 another market opened, provided by the railway: the needs of the fast-growing cities, homesteads and farms of the prairie wheat boom. Dominated by national railways, working in different types of forests, isolated from the coast geographically and culturally and facing not one but two sets of government rules (federal and provincial), the interior industry developed a sharp-edged, independent spirit and a distrust of coastal policy right from the start. This distinctive historical difference runs deeply through British Columbia's forest history and reverberates even today, just as the interior dominates the forest economy of the province.

Growth and Debt and Debate

The arrival of the railway in 1885 also brought a major shift in provincial regulation. Leases dominated the supply of timber for the large mills. To protect the smaller logging operations, the hand-logger's licence was created, allowing access to Crown land as long as steam tools were not used. Amendments to the *Land Act* in 1887 saw transferrable leases awarded by bid on the coast (favoured by large-scale investors and mills) and the creation of the small, one-year special timber lease, which allowed the smaller interior operator access to Crown timber. Historians Peter Gillis and Thomas Roach argue this is the moment when the provincial government began to see timber as a promising revenue source for the future.

By the 1900s a great population surge into the Canadian west was underway. It wasn't just the interior industry that prospered from prairie demand for lumber. The coastal industry also took advantage of the frantic prairie demand for lumber, despite spot markets like the demand for lumber to rebuild San Francisco after the 1906 earthquake and fire. Estimates are that two-thirds to three-quarters of British Columbia's production went east to the prairies between 1901 and World War I.[14] The profitable eastern markets offered

Logging with oxen on an early corduroy log road. Brunette Sawmill Company, Nicomikl River, about 1870. C-00570.

by railways and farming drew attention away from developing or keeping overseas foreign markets, a constant problem for each early chief forester.

The arrival of so many immigrants to British Columbia at the turn of the century was similar to the influx during the gold rush. It, too, created a costly infrastructure problem for government, as new towns, railways and roads were needed. The agrarian and resource frontier placed strains on the fiscal resources of government to manage growth at exactly the same time that investment capital was pouring in and dramatically distorting BC's forest industry. Those already at work in the forest industry were equally upset about the pace of change and wanted their regional views to also be heard. All of this came together in a historical moment, a great crisis that brought about a public debate over the future of BC's forests. This debate would be resolved by an act which created what we know today as the British Columbia Forest Service.

Chapter 1
The Great Forestry Debate, 1906–1912

By 1906 it was becoming clear to many British Columbians that the provincial forest industry had grown to such a scale that 19th-century settler-based management policies were no longer adequate. Led by Richard McBride, the first BC-born premier, the government was recognizing that there were problems in many parts of the industry, as well as fiscal problems for government. No one was happy with the situation.

The social debate over how to manage British Columbia's diverse forests had many root causes. On the coast, loggers were unhappy with the price set by the large sawmills. They saw a simple solution in the easy, water-borne transportation to the south: sell their logs to the mills of Puget Sound. Independent logger-operators began to challenge the existing regulations. A flaw in legislation banned the export of logs from leases and special timber

Allison Camp and Shingle Mill at Greene Point Rapids, 1912. H-00426.

Scaling logs on Vancouver Island, about 1900. B-07327.

licences but said nothing about private land or hand-loggers. The independent contractors on the coast began to exploit that loophole, seeking to bypass low mill prices and export to Puget Sound's American mills. The provincial timber inspector began to suspect that logs from leases were ending up in the booms of private or hand-logger operations. With only a single inspection tug on the coast, monitoring and policing were impossible. The potential for log smuggling and lost revenue worried the timber inspector.

From 1904 to 1906 the British Columbia Logger's Association president, J.S. Emerson, fought a series of legal skirmishes with the government over log prices and exports. When his boom was seized by Timber Inspector Robert Skinner, assisted by sheriffs, Emerson challenged the government in court, won, and sought damages, selling some of his boom locally and exporting the rest. The uproar cost Skinner his job, and Premier McBride moved quickly to close the loophole, appointing John Murray to regain control of the situation. This incident convinced the provincial government of the need to control log scaling in order to stop smuggling to Puget Sound's mills. Established by the *Timber Measurement Act* of 1902 (replaced 1906), the newly appointed Supervisor of Scalers, Andrew Haslam, ordered his staff of government scalers to expand operations to include grading logs, something previously left to the mills.[1]

The discord over log exports was also set against the background of increasing US interest in investing in BC's forests. After the first great rush of capital into the Pacific Northwest, industry players in the south were becoming uneasy. The US government began to withdraw western public lands from sale in 1891 and placed them into forest reserves, later known as the National Forest System (so named in 1900). In 1897 alone they created over a dozen such reserves, all to serve national strategic interests – a fact often forgotten in the rhetoric of free enterprise. When President Theodore Roosevelt left office in 1907, he placed 16 million acres (6.5 million hectares) of western lands into the National Forests.

The realization spread among American lumbermen that even in the Pacific Northwest, timber supplies were no longer limitless. As a result, the value of standing timber, such as Douglas-fir, began to climb, and new pressures on access were created as the capital-intensive pulp and paper industry emerged and also began looking for secure supplies. For example, shrewd large operators such as Frederick Weyerhaeuser invested in large purchases of railway syndicate lands to obtain stable, long-term sources of timber. Control over timber also became an important factor as collateral to entice investments of the large amounts of capital needed to build industrial paper mills.

Scaler Allen H. McLean stamping logs, 1914. NA-04455.

The Conservative government of Richard McBride seized the opportunity to dramatically increase revenues from forests by modifying the terms of transferable leases and allowing the extension and transfer of the special timber licence. And so, in 1905, leases were folded into the licensing process, with its ground rent and higher royalties, and tenure was extended from five years to 21 years. Made transferable and with no limit on the number a person or company could own, licences became commodities. One company held 375 licences and individuals buying 100 to 200 licences were not unknown.[2] In 1905 there were 1451 licences in use, but by the moratorium of 1907, a total of 15,160 valid licences had been issued.

The changes in legislation and the perceived North American timber shortage brought a flood of investment and speculation to the forests of British Columbia. Government immediately benefited, as 40 per cent of provincial revenue was provided by the licences. This not only dealt with a $10 million debt but provided a steady flow of revenue to undertake large developments. With a steady income stream government could use to fund development, and alarmed at the massive buy-up, a moratorium on new licences was unexpectedly declared on December 24, 1907, by an order-in-council which reserved all unalienated Crown timber. Having achieved its revenue goals, the closure also served to protect industry

Scaler Evan R. Thomas scaling a raft of logs, 1914. NA-04456.

from overproduction and allowed government to reopen the gates if a combine of licence holders grew too strong or stimulus was needed.

One other unexpected event had driven the last wave of speculation just before the closure of licences. At 5:12 am on April 18, 1906, 80 per cent of the city of San Francisco was destroyed by an earthquake whose magnitude is still subject to debate. It was followed by several days of fires, and at least a quarter of a million people were homeless. There was wild speculation in British Columbia and in every timber- and lumber-producing region in North America to supply the massive rebuilding effort. Inevitably, prices in San Francisco collapsed as inventories grew, creating a different problem for BC producers and underfinanced speculators alike. The massive speculation did drive up stumpage costs for loggers. It flooded the market with lumber, and then crashed the price for logs as demand from quake-damaged San Francisco inevitably slowed. The BC Loggers Association demanded that speculators who tied up large amounts of timber be

THE DOG-IN-THE-MANGER LOGGER.

PREMIER McBRIDE:—"So you won't develop the country yourself, and won't let anyone else come in and do it, eh! We've too many of your sort here, my fine fellow. Out you go, quick, and make room for Capital and Industry.—*See page 4.*

Premier Richard McBride, supported by British investment capital, whips BC loggers into submission in this political cartoon by Emily Carr. *The Week*, June 10, 1905.

Courtesy of Dr Patricia Roy.

driven out, the new licence holders demanded action by government to protect their investments, and all battled for political and public support in the press.

Over the winter of 1907–08 the industry was divided between the views of loggers, coastal mill owners, new investors (led by Ontario's former provincial forester Judson F. Clark) and the older Interior Operators Association who wanted the security of longer-tenure terms. McBride's government also faced the problem of re-election. How to resolve the situation was a pressing question, and the intensity of the debate weighed heavily on McBride. His decision was to bring everyone together to air their views in the 1909 Royal Commission on Timber and Forestry.

The 1909 Royal Commission on Timber and Forestry

In July 1909 McBride's Conservative government appointed a three-person Royal Commission to launch public hearings and make recommendations about timber and forestry. The commission members were the Hon. Frederick John Fulton, a staunch member of McBride's government (elected from Kamloops) and also chief commissioner of lands (responsible for forest licences); Arthur Samuel Goodeve, the member from Rossland; and Alfred Cornelius Flumerfelt, who had been the president of Hastings Shingle Manufacturing, one of the

Alfred Cornelius Flumerfelt, 1910.
A-02151.

largest shingle mills in the province. Flumerfelt was also a member of the executive of the BC Lumber, Logging and Forestry Association. Legal counsel for the commission was James A. Harvey, KC, who was also active in lobbying and promoting the interests of the Mountain Lumber Manufacturers and their demand for longer tenure.

Over the next 13 months the commission visited a dozen towns and cities to hear testimony from 116 witnesses. In a statistical study of the evidence before the commission, Robert Marris found that businessmen made up two-thirds of the witnesses and government a further quarter. Only four people testified on behalf of organized labour.[3] Most spoke on behalf of hand-loggers and against perpetual licence renewal, which they felt would be abused and lead to an eventual erosion of government control of Crown lands. Yet none of this pro-business stance should be surprising in the context of the times. Clearly, the commission was seeking to find a compromise between regional industry and business factions, and trying to find a balance in tenure options that would benefit government, large and small operators, and serve the interior and coastal regions.

To see the commission as an exercise wherein big business and government attempted to carve up the forests is to discredit a very real and deep concern of the commission: what should British Columbia's forest policy be in the long term? Their concern was genuine. The commission was also established at a time when Canada and the USA were both engaged in national discussion about the meaning of conservation of natural resources.

The commission heard evidence on a wide range of issues, including how much is a fair return to government for timber removed from Crown land, who should pay what proportion of firefighting costs, how many licences are too many, and what should be the relationship between cutting and sustaining a forest. As each special interest presented its own interpretation of what was needed for good policy, it became apparent that what was missing was reliable information on British Columbia's forests – information critical for making plans for the forests' future.

Forest tenure and how to reform royalties to take into account future value became a significant focus of the Royal Commission on Forestry. A comparison of the transcripts of the hearings shows that the discussion revolved around tenure and regulation, as well as the views of different factions and the coastal and interior industries. The commission firmly rejected the idea of the privatization of forest lands as being not in the best interest of the public or government. In British Columbia the development-minded government of Richard McBride kept control of 43 million hectares, believing, like their American counterparts, in the public forest.

Frederick John Fulton.
Kamloops Museum and Archives 387.

One of the first items of evidence in the Fulton Commission papers is a transcript of a speech given at the First National Conservation Congress in Seattle on August 27, 1909, by Gifford Pinchot, the fourth chief of the Division of Forestry and first chief forester of the US Forest Service.[4] While there were few ecological questions asked about forests, there was a fledgling discussion about the possibility offered by nurseries and seed trees for establishing a permanent cycle of regrowth and second cuts.

Fire dominated much of the discussion relating to settlers. For the commission, the problem was the use of fire to clear land. It was argued that while $55 million worth of timber had been cut in the Crowsnest Pass area of southeastern BC, over a half a billion dollars worth had been lost to fire. The evidence presented to the commission on the fires of 1909 singled out settlers and sparks from steam engines as primary culprits, not nature.

Table 1. Fire in 1909: Known Causes of 203 Forest Fires. Source: Royal Commission on Timber and Forestry, 1909. BC Archives, GR271, Box 1, File 8, F24.[5]

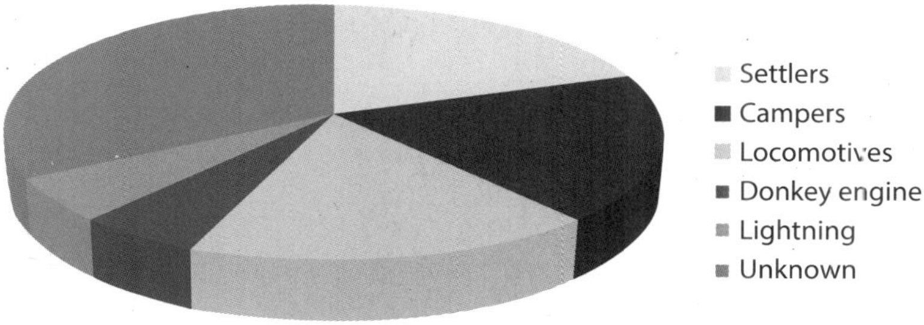

Suggested controls on settlers' use of fire land ranged from Mr A. Leitch's suggestion that "Isolated settlers should not be allowed among standing timber" to a ban on open fires from mid June to mid September. At the hearings, several endorsed a government burning permit system.[6] The testimony of a future premier, the Hon. John Oliver, Member of the Legislative Assembly for Delta, cut to the heart of the matter. Unlike many of the other witnesses, he spoke as a

Slash burning by settlers along the Canadian Northern Pacific Railway, 1913. NA-03725.

representative of farmers and as a Liberal addressing a commission reporting to the Conservative government. Oliver pointed out:

> my experience in this country is that it is impracticable to clear land except in summer, especially on the coast, and if you are going to prohibit the setting of fires except in certain months, you are practically prohibiting clearing on any large scale. Of course it is possible in winter to blaze it and pile, but it more than doubles and trebles the cost of clearing and you practically prohibit the settlement of timber lands.[7]

Oliver pointed directly to the problem of supporting agricultural land clearances and timber-cutting licences. The old system of timber giving way to a settlement frontier was no longer working. The limits of the forest were becoming clear to industry in coastal and southern BC. Oliver added:

> A great deal of evidence has been given before the Commission regarding lands and the danger of fire. I want to say to you frankly it is a case where you cannot eat your cake and have it. You cannot protect your timber against fire and yet clear up your land. It is a question of conflicting interest, where interests are opposed one to the other.... You have got a serious question to consider, whether you are going to make the interests of the timberman superior to the settler.[8]

Oliver acknowledged that settlers were already restricted from entering any land capable of producing 8000 board feet of lumber per acre west of the Cascades and 5000 to the east.[9] He favoured no homesteading on licensed lands but suggested that the proper function of government was to mediate between the settler and the licence holder.

> Q. Who should determine; should it be in the hands of the government?
>
> A. Certainly. It cannot be in anybody else's hands. They are responsible and must stand between the holder and the man who wants to clear.[10]

The commission puzzled over the separation of the problem of settlers and industrial forest tenure. It was both a practical conflict and a political problem for McBride. One could argue that here began the discussion of what in 2014 we call the wildland-urban interface issue – fire protection at those locations

A farm at the edge of the forest, 1905. I-56454.

where towns and subdivisions meet forest and dry rangelands, the most memorable being the Okanagan Mountain Park Fire near Kelowna in 2003, one of many such incidents.

The solution to the problem of the contested forest was far from perfect, but from the start the idea began to emerge of multiple-use forestry. It is no coincidence that at the same time as the commission set out its recommendations for forestry, we find the ideals of conservation in the creation of British Columbia's large parks, such as Strathcona (1911) and Mount Robson (1913). Policy began to move from special economic interests to the concept of protected areas: Crown lands reserved for neither settlers nor industry, but for something called "the public."

The Royal Commission on Timber and Forestry has its own complex history, but its recommendations were widely read and well received. This theme of policy for the public good was front and centre in a remarkable speech presented to the legislature by Minister Ross at the second reading of the 1912 *Forest Act*, which came out of the commission's recommendations. Whether the commission's secretary, Martin Grainger (a future chief forester of BC), had a

Staff caricatures have a long tradition. M.A. Grainger; H-05548.

hand in its creation is unknown. Ross represented Fernie, a town destroyed by fire in 1904 and subsequently rebuilt, and he knew well the regional economic importance of the forest resource.[11]

"A Common-Sense Conservation Policy"

> An Epoch, sir, is drawing to a close – the epoch of reckless devastation of the natural resources with which we the people of this fair young Province have been endowed by Providence – those magnificent resources of which the members of this Government and this Assembly are but the temporary trustees. That rugged, rudimentary phase of pioneer activity is doomed to end. The writing is on the wall; the writing – to put the simple fact – is in this Forest Bill. Armed with that weapon, as forged by this Honourable Assembly, the Government of British Columbia will undertake the work of forest conservation.

> Sir, a certain solemnity of its own surrounds the introduction of this forest bill ... we call the world to witness that we legislate today, not only ourselves and the needs of this day and this generation, but also, and no less, for our children's children, and all prosperity – that we may hand down to them their vast heritage of forest wealth, unexhausted and unimpaired[12].

In his speech introducing the new *Forest Act*, Minister Ross reviewed the work of the Forest Commission and the tremendous changes he saw underway in the previous decade in the province. McBride's government understood well that the rapidly growing province meant tremendous costs, with investments in railway expansion, growing cities and agriculture's need for irrigation, land and roads. British Columbia's ports and harbours were investing heavily for the forthcoming opening of the Panama Canal, expected in 1914. This new canal promised lower costs and more rapid connections to European and eastern markets for the new pulp and paper industry and lumber. The act was intended to resolve the government's financial needs for the province's building boom,

The Honourable William Rocerick Ross, K.C.
ZZ-95143.

and it had every indication of success. In 1904 the province received $455,000 in revenues from forestry, a seventh of its income. In 1908 revenue was up 500 per cent to $2,785,000. By the time the bill was being debated, almost $13 million had flowed into the government's treasury and forestry was king, providing 40 per cent of provincial revenue. The problem was how to stabilize this windfall revenue for the long run.

Ross referred directly to the problems in eastern Canada and the eastern United States, where cut-over lands and declining revenues faced once-prosperous forest-based economies. A heavy taxation burden would harm local producers and drive away much needed investors, but too-lax taxation would give away the province's future wealth and lead to the crashing economic fallout of overproduction. He spoke as well about the clear evidence of limited forest resources in the decline of the forest industry in Michigan and Minnesota and the end of eastern Canada's White Pine industry. There was no clearer sign of limits than President Roosevelt's decision, advised by Gifford Pinchot, to commit to conservation of resources by removing 170,000,000 acres (68 796 559.181 hectares) of forests into a National Forests reserve. The importance of the industry's westward shift was clear to Ross. He pointed out that half of the United States' forests were in the Rockies and on the Pacific coast, and British Columbia held the same proportion of Canada's forests.[13]

Anxious to attract capital for pulp and paper or general investments, but wary of the all too easy trap of oversupply, the government of the day sought a secure and permanent flow of revenue that would also balance the needs of investors and producers, both large and small. But it would be wrong to think that only fiscal issues dominated the philosophy behind the first *Forest Act*. The bill sought "a common-sense conservation policy," according to Ross.

> Without such a policy, the cutting of the present crop of timber would ruin and lay waste our timber lands. The young growth would be burned, inferior species would replace the Douglas-fir; by the denudation of our watersheds soil erosion would take place on mountain slopes, irrigation would be endangered, the lumbering industry would gradually decline, and British Columbia would sink into the stagnant insignificance that has overtaken other worked out forest regions. That dismal fate, however, should never overtake this Province, because we nowadays know how to avert it, and because the Government of this Province would see that necessary and essential measures were taken. This government would do this now – not when all would be too late.[14]

Implementing Progressive Forestry

A strong influence on the act and the commission was the progressive era's strong belief in supporting not only new institutions and ideas but also the young professionals of the new generation. The late 19th century saw an amazing revolution and hybridization that created scientific forestry. North American students like Gifford Pinchot were exposed to differing French and German national schools of forestry, with their military traditions and emphasis on fieldwork and sample plots. Traditional forestry was itself transformed by the new biological sciences that led to the concept of ecology, and also by the knowledge gained from the world's colonial forests. Young German foresters such as Sir Wilhelm (William) Schlich and Sir Dietrich Brandis created textbooks comparing European forests with others they had worked in – in Africa, Southeast Asia and, most influential of all, India's Forest Conservancy system, which gave North American foresters the concept of resource conservation. Forestry became a global idea in which there was regional diversity.

The first American Forestry Congress was held in two locations in 1882, in Cincinnati, Ohio (April) and in Montreal (August) in a continental discussion as these new ideas met head-on with policy and industry concerns over limits

Natural regeneration, Campbell River Experimental Forest, 1931. NA-05011.

and proper conservation of resources for future generations. New schools and practices about forestry emerged at places like Biltmore, Yale, Cornell and Toronto, with leading thinkers like Pinchot, Schlich, Brandis, Carl Schenck and Bernhard Fernow. The new ideas also gave rise to new forestry schools, which graduated the first wave of North American-trained professionals, such as the first grads from the Yale School of Forestry in 1902 and, in Canada, from the University of Toronto's School of Forestry, founded in 1907. Canada's Commission of Conservation struck a committee in 1909 to look at the future of Canada's forests, and in 1911 they held a major conference to discuss progressive forestry in Canada. That same year saw the passing of a *Forest Reserve and Parks Act* for BC and the introduction of high-lead logging, which would accelerate coastal logging.

To pursue good forest stewardship, British Columbia's government proposed to create a British Columbia Forest Branch. The Fulton Commission members had consulted thoroughly with Gifford Pinchot, the US Chief Forester. Pinchot's assistant in setting up the United States Forest Service was Overton W. Price of the National Conservation Association, who served as a consulting

Logging Camp, Montana Lake, near Revelstoke, 1912. A-09444.

advisor to the commission. Pinchot was ever the great forester-as-politician, but Price had a true organizational mind. As James Lewis has pointed out, it was Price who gave the US Forest Service a clear and decisive structure. He also created a structure which carried out Pinchot's wish for decentralized decision-making, to give rangers and districts the power to make decisions locally, on the ground.[15] For the British Columbia Forest Branch, that principle has its origins in Price's advice to the first Royal Commission.

Organizational Roots of the Forest Service

The proposed BC Forest Branch would also bring together several other pre-existing groups, each on their own separate budget vote that existed within the Department of Lands and who were already at work on different aspects of the forest industry. One of the reasons that Chief Forester Harvey R. MacMillan noted for previously having separation between the Timber Inspection Branch, which collected royalties, and the scalers who measured logs was to have each act as a check on the other. Then there was the fire warden system, based on a five-month, seasonally employed workforce.

The Timber Inspection Branch of the Department of Lands had evolved out of the creation of the provincial timber inspector, an office created under the 1888 *Land Act*. Originally a three-person Vancouver office collecting royalties, its work expanded. Clerical staff, assistant timber inspectors and early forest rangers inspected pre-emptions, issued hand-logger licences and dealt with applications for timber. The branch had three patrol launches: the *R.J. Skinner*, which worked 700 miles (1126 km) from Sayward to Prince Rupert, north of Jervis Inlet; the *Kiora*, which patrolled eastern Vancouver Island and the Gulf Islands; and the *Alanbee*, which patrolled New Westminster district and islands.[16] After the 1906 requirement for in-province manufacturing of logs, they also kept a watch out for illegal log exports along the boundary waters above Puget Sound. Assistant timber inspectors worked out of government offices in Nelson (1897), Kaslo (1904) and Kamloops (1909). Just before the Forest Branch was established, they were responsible for monitoring 350 logging camps with 8000 men, as well as 250 sawmills with 7500 employees.

Scaling timber for stumpage royalties was originally conducted by the mill owners themselves, before giving way to licensed scalers. The first rules implemented were from *Scribner's Lumber and Log Book* (1882). Eastern Canada quickly adopted it, with the inclusion of Doyle's table on sawn lumber dimensions and the ever useful wage tables for sawmill payrolls. However the characteristics of Pacific species, such as the flared butt and taper of Western Redcedar, differed from rules designed for eastern Canadian species.

Horse drawn logging, Bonney's Ferry Lumber Company, 1913. NA-03494.

New Westminster's Mathew L. Foley created the *British Columbia Log Scale* (1891), the recommended choice of the earlier Royal Commission on Scaling (1894–95).[17] Foley's rule emerged as the BC Provincial Log Scale in 1895, the provincial standard. Mill owners were unhappy at the cost and problems of accounting for lumber acquired from different subcontractors at different leases and limits. Government was uneasy about logs being cut along the coast and disappearing across the border to mills in Puget Sound, with royalties calculated only on the logs received at the mill. The solution for both was for government to take responsibility for scaling, billing five cents per 1000 board feet measured.

The office of the supervisor of scalers had been created by the provincial government in 1906 to deal with the industry's rapid expansion. Four years later, an assistant supervisor of scalers, based in Nelson, BC, took over supervision of scaling in the southern interior, from the eastern Cascades up to the Grand Trunk Pacific Railway line being built from Prince Rupert to Prince George and out to Edmonton. Government log scalers were stationed at Nelson, Victoria, Nanaimo, McRae Cove, New Westminster, Ocean Falls and Vancouver, with acting staff in the Kootenays. With the creation of the Forest Branch, the Scaling Branch also merged into the new Timber Management Branch.

Before the *Forest Act*, fire protection was provided by the Protection Branch. Based in Vancouver, it was a five-month, seasonal operation with no permanent field staff. The 1888 *Land Act* provided for the appointment of fire wardens as

Bunch Grass, range patrol, Marry Creek, 1914. NA-04114.

needed, but it was not until 1905 that fire wardens were hired to pre-emptively enforce the *Bush Fire Act* during the fire season. The expansion of the forest industry, new settlement and more railway traffic, coupled with dry seasons and lightning strikes, led to the creation of the Office of Chief Provincial Fire Warden (1907), and by 1909 there were 37 seasonal wardens at work across the province. By 1911's fire season the province had 110 fire wardens and six marine patrol vessels. Comments made just prior to the *Forest Act* called for more permanent improvements:

> The wandering patrolman, who is sometimes hard to find when wanted in an emergency, has had his efficiency vastly increased in many sections of the West by the establishment of field-telephone systems to which he may connect his portable line and instrument; by the cutting-out of fire lines and trails; by signal systems worked from lookout stations placed in commanding situations, and in many other ways.[18]

The year after the Royal Commission's final report was published, the chief provincial fire warden's staff included a supervisor for the coast division and divisional fire wardens at Victoria, Vancouver, Kitsalas and Barkerville. The supervisor of the mountain division (which included the Cascades) had divisional fire wardens based in Grand Forks, Fernie, Cranbrook, Vernon, Revelstoke and Nelson. Each division had patrol districts that were staffed only during times of fire danger. In other years, wardens were posted at Port Simpson, Cowichan, Trout Lake, Atlin, Nicola, Comox, Hedley, Slocan, Quesnel, Heriot Bay, Mission, Bella Coola, Howe Sound, Chilliwack, Tête Jaune Cache, Golden and many more locations. Vancouver alone had 20 patrol districts. On April 5, 1912, the Forest Protection Branch was combined with the new Forest Branch, with R.J. Benedict as chief of operations.

In 1913 as the first Chief Forester H.R. MacMillan would bring all of these earlier Department of Lands branches – Timber Inspection, Scaling and Fire Protection – together into the new Forest Branch. He favoured districts that combined services and were run by one carefully selected individual, the district forester. Each district forester would learn their area and coordinate services, eliminating inter-branch rivalries and reducing expenses.

Forest assistants would be assigned temporarily to seasonal work in a district, performing land classification, surveying and timber cruising. Head rangers would take responsibility for fieldwork in large sections within a district, reporting to the district forester. District foresters, in turn, would supervise rangers, who were

On patrol in the Sheep Creek, Windermere area, 1914. NA-04614.

stationed to supervise the timber cutting, issue grazing and burning permits and oversee temporary staff on fire protection duties. All this would be supported by a centralized administrative staff in Victoria who would deal with mapping, licence records and assembling annual data for planning and reports. The centralization of clerical staff from all of these earlier activities was a major task in itself, involving the standardization of over 50 types of forms and account books – a figure only covering existing activities and not including any attempt at a provincial forest atlas or inventory. Consolidating these previous areas with BC's new Forest Branch also provided a core of experienced staff familiar with the areas and the communities in which they worked. (That said, MacMillan noted that he had in mind the cutoff age for a ranger as 40 years old.)

The formal structure decided upon for the new Forest Branch ended up being more ambitious than the original recommendations of the Fulton Commission. The Minister of Lands asked Overton Price, who had helped establish the organizational structure of the US Forest Service, to act as consulting forester during the 18 months of reorganization and creation of the new Forest Branch within the Department of Lands.

Chapter 2
The Forest Branch, 1912–1920

Harvey Reginald MacMillan (1885–1976) was a second-generation Canadian from a Scottish family in Ontario. He had a difficult start in life, first with the death of his father and the extreme poverty his mother endured. Tuberculosis took a toll in his family. He trained at Guelph in agriculture and, determined to avoid farming, developed an interest in the new forestry movement. In 1904 he worked summers for the new federal Dominion Forest Service, which had been established only five years previously. MacMillan worked first at the Indian Head nursery in Saskatchewan, and in 1906 he conducted surveys at the Forest Reserve at Riding Mountain, Manitoba, and wrote his thesis on its problems. With a University of Toronto degree for his Guelph studies, he went on to the

Staff caricature of a young H.R. (Harvey Reginald) MacMillan, CC CBE. NA-22319.

two-year Master's program at Yale's Forestry School. It was during that program that he made his first trip to British Columbia, doing timber cruises at Powell River at the height of the licensing boom in 1907. He gave the graduation address for the Yale forestry class of 1908 and went to work for the Dominion government surveying forests that were to become national parks, such as Waterton Lakes in Alberta.

That fall, what he called "The White Plague" – tuberculosis, which had taken such a grim toll in his family – struck him again. For the next 30 months, the man who became Canada's most famous forester was bedridden. He spent the time reading history, corresponding, and contemplating forestry

The first staff of the Forest Branch, outside Annex Building at the corner of Superior and Government streets, Victoria, April 1, 1913.

Front row (left to right): Percy Le Mare (District Forester, Lillooet), R.E. Benedict (H.Q.), A.M.O. Gold (Foresters Assistant, Cruiser), J.R. Martin (District Forester, Nelson), Walter L. Loveland (Ranger, Vancouver Island), M.A. Grainger (Asst. Chief Forester), S.W. Barclay (Royalty Inspector), P.Z. Caverhill (District Forester, Kamloops), J. G. Sutherland (For. Asst., Cruiser); R.L. Campbell (Librarian and Publicity), Thomas Mouat (Acting Superintendent of Scalers), H.R. MacMillan (Chief Forester), J.D. Gilmour (District Forester, Cranbrook), H.R Christie (H.Q.), A.K. Shives (For. Asst.), H.G. Marvin (District Forester, Ft. George);

Back Row (left to right): W.C. Gladwin (H.Q. Records), H.K. Robinson (District Forester, Vancouver Island and Operations Chief of Surveys), W.J. Van Dusen (Asst. Chief, Management), Jack Stilwell (Vancouver Island Patrols); L.R. Andrews (District Forester, Vernon); P.S. Bonney (For. Asst.), H.C. Kinghorn (For. Asst.), C. MacFayden (District Forester, Tete Jeune Cache), H.S. (Buck) Irwin (Acting District Forester, Prince Rupert), H.B. Murray (For. Asst , Kamloops), G.H. Prince (For. Asst.).

In the window: S. Hawkins (Office Boy, Water Rights Branch).

N. Brown photograph; RBCM research, private collection.

conservation issues and the problem of sustainable resources. In the summer of 1911 he was discharged from the sanitarium, returned to work at the Forestry Branch in Ottawa and married Edna Mulloy. In July 1912 he accepted the offer of a position as the first chief forester and head of the new BC Forest Branch.

In his new role, MacMillan used his extensive forestry connections to gather around him 26 of the brightest people he knew, including old survey party friends like Whitford Julian Van Dusen, who he first met through Bernhard Fernow.

Chief Forester H.R. MacMillan in His Own Words

As it happened when I got there ... they thought they'd hired the wrong fellow. I was too young. But I fitted the specifications that they had to have a trained forester and he had to be a Canadian. It happened, that particular year, 1912; I was the only one that could say the specifications fit me. I got the job, but when they looked at me, they wished to God they hadn't got me. I was too young and fragile and unwise. But the Minister was too proud to fire me. He waited. I think that is a habit of Ministers. He waited until I had made a mistake, and I made a good many, but he didn't catch me at them.[1]

The new Forest Branch was organized into 11 districts, based in Cranbrook, Hazelton, Kamloops, Lillooet, Nelson, Prince Rupert, South Fort George, Tête Jaune Cache, Vancouver, Vernon and Victoria. Each would have a district

District forester headquarters at Tête Jaune Cache, about 1912. NA-03619.

Miss A. Hughes, Forest Service librarian, 1949. NA-09777.

forester, and they would have on average three rangers, each with a ranger district. Other staff working out of the ranger's station included a forest assistant or technical assistant, who dealt with timber cruising, reconnaissance surveys and silviculture. There would be timber scalers and temporary seasonal forest guards work on patrols, summer firefighting and improvement projects. Clerical staff would assist at all levels, and mapping would take place first at the ranger station. Maps, correspondence, permits and reports would flow up to the district and headquarters clerical staff in Victoria. Victoria would produce better maps, inventory data and instructions based on what the station and district supplied and dispatch special survey teams for areas needing extensive work.

In Victoria the headquarters unit consisted of the chief forester and his clerical staff. Reporting directly to him were four assistant foresters, each with their own special management area.

Significant additions to the timber management section were the expansion of protection to include firefighting and responsibility for grazing and range management. The timber management section of the branch assembled the forest atlas, dealt with land classification, scaling, log seizures and trespass, silviculture, marking and inspection, timber sales and anything related to lumber production. Operations began with organizational structure and personnel issues, education and publicity, and information services such as a small library and forest protection.

The records section dealt with appointments, revenues and supervising expenditures. They were also responsible for determining the status of land, sales and royalty work, drafting new legislation and the collection of statistical information.

The timber surveys section undertook the comprehensive surveying of the province's forests, a task that in 1912, when he introduced the first *Forest Act,* Minister Ross hoped to see completed by 1918. This was the branch hit hardest by the onset of World War I, as many of its rugged, young, self-reliant staff responded immediately to the call for volunteers.

No sooner had the new Forest Branch begun its duties when the economic depression of 1913 stalled the forest industry and impacted government budgets. With district offices only slowly being established, headquarters launched much of the initial survey work and then transferred it to the districts as they were able. Headquarters also provided educational publications for settlers that were designed to help get people established and minimize conflict between farm and forest.

Forest service staff transporting a canoe on the 27-km portage between the east end of Ootsa Lake to West Cheslatta Lake, 1928. H-03116.

MacMillan saw in the previous organization a duplication of resources and a compartmentalization of knowledge among the different branches that might give rise to jealousies and conflict. He favoured consolidation through single authorities, be they himself, his assistants or the district managers and their rangers.

> This system which naturally develops men of all-round ability and experience, and affords a much greater opportunity for rapid advancement than one founded on the specialist plan, is a very important factor in attracting and holding good men in the service. Another advantage over the specialist plan is that it is easily understood by the public, since all business relating to forest administration and protection is transacted by one and the same officer.[2]

He also argued for a permanent protection staff, kept busy doing surveys and trail work outside the fire season. He believed this would attract and hold a better quality of staff than temporary fire season workers.

MacMillan set out a philosophy for the Forest Branch that guided many staff over the years, one centred on cooperation.

The new Forest Branch administered vast forests, like this one dominated by Sitka Spruce. NA-06522.

MacMillan wrote:

> The Forest Branch is young, and it faces a great opportunity. The material welfare of British Columbia is inseparably bound up with the success of her forest policy. In the last analysis, the success of the Forest Branch will depend in large part upon the knowledge, and consequently the understanding, which the people of British Columbia have of its work.
>
> The Forest Branch is dealing administratively with a public resource – a resource which is the property of the people – and the Branch is, in a very real sense, the steward of the people and the servant of the people in the protection and use of the people's own property. Through its bulletins, through addresses, by personal intercourse, and above all by a frank and direct method of procedure, the Forest Branch will do all in its power to inform the people of British Columbia about their own property which the Forest Branch administers in trust. It will also make a special effort to obtain in their own and in the common interest the co-operation of lumbermen and other forest-users in its work.[3]

Logging aircraft spruce in Haida Gwaii during World War I. D-04662.

That definition of "the people" in 1912, however, reflected its times and did not include the 16 companies that employed "Oriental labour" to cut timber. Staff also served notices for them to come into compliance with the anti-Oriental labour clauses of their licences.[4]

MacMillan held a philosophy about the importance of the Forest Branch being steward of the people's forests. His first report as chief forester (dated January 7, 1913) argued that the forest was growing faster than the industry, and his solution was to grow the industry to match that production and to recover the wood he saw as wasted by natural decay. His estimate was possibly three billion board feet of timber growing on one hundred million acres. If industry was only capable of cutting 1.25 billion board feet a year, it would take 250 years to cut just the standing mature timber.

> To sum up, then British Columbia possesses forests whose actual production in wood each year is alone five times her total lumber cut. What is not cut is wasted in the end. It is not merely advisable to encourage the growth of our lumber industry until it equals the production of our forests – it is our clear duty to do so, in order that timber which otherwise will soon rot on the ground may furnish the basis for industry, for reasonable profits to operator and Government, for home building, and in the last analysis, for the growth of British Columbians.[5]

Creating Revenue and Information Services

If the districts were to act successfully with public confidence in the Forest Branch, it also needed sound planning, and that relied upon a good flow of information in from the field. Good community relations also meant removing friction points, such as revenue collection, and managing the increased revenues required clear accounting practices with checks and balances. Putting all of that in place required an effective and efficient headquarters administration – one of the key lessons from Price's work setting up the US Forest Service and the example of the Dominion Forest Service in the railway belt. MacMillan moved on that issue first before tackling fire protection.

In 1900 all the paperwork for British Columbia's forests was handled by a single official. That person handled the occasional timber lease, received rent and conveyances on some 220 other leases, issued timber marks and gave out licences to cut on vacant Crown lands. The 1905 licensing boom quadrupled the official staff, and from that point on, each subgroup grew in complexity, in some cases duplicating what another was doing. The work grew with the scale of revenues generated, the number of accounts, the return of surrendered leases in economic downturns or after unfavourable cruising reports. Managing fire season staff, supply expenses, personnel appointments – all these put more pressure on financial accounts.

The end of the Forest Branch's first fire season brought little relief to the new administrative staff, as the Convention of the Canadian Forestry Association arrived in Victoria in September to examine the new organization and enjoy BC's forests. As MacMillan put it, it was not until the end of October that the "wanderings of the scattered staff from one committee-room and basement to another were brought to an end by completion of the present office building."[6] By the end of November the staff, filing cabinets, desks, records and all the different types of forms and account ledgers all were merged under one roof.

The head office quickly established specialties for monitoring leases, determining the status of lands, setting out survey objectives (until the work could be transferred to the new districts), and handling the accounts, expenses and revenues – much to the relief of fieldworkers who were freed from office tasks that kept them from the forests.

A System for Fire Protection

MacMillan had made the decision at first to leave the fire protection system – an "imperfect instrument" – alone while he concentrated on other areas of the new organization. He was lucky. The summer of 1912, unlike 1910, had a very quiet fire season, with only 160,000 acres (65,000 ha) burned over. Of 347 fires in 1912, half had known sources. The leading culprits were controlled burns of logging slash set too early in the season, followed by fires caused by campers. MacMillan noted with satisfaction that only eight of the 9400 fires under permits had escaped. There were 11 convictions for fires without permits. At logging camps cloth notices were posted about procedures for fire protection around donkey engines. Some 300 donkey engines were inspected to check for their spark-arresters, fire gear, water, hoses, and ash and coal disposal procedures. Inspection of railway locomotives was a Dominion jurisdiction, and liaison work with the Dominion officials, railways and logging companies formed another part of fire protection.

MacMillan agreed with his predecessors in protection that trail work and telephone systems were the key to fast response. The typical telephone trail of the time ran 40–75 miles (65–120 km) across isolated areas, connecting districts with rangers' cabins. Fast communications and response were the keys to effective fire control.

The other key to effective response was the lookout system. Trails and telephone lines connected a series of proposed lookout towers to be established at key vantage points around major valleys, rivers and lake systems. In 1913 district foresters began producing plans for fire protection. In the Okanagan it was L.R. Anderson of the Vernon forest district and District Forester J.D. Gilmour of the Cranbrook forest district who produced the first comprehensive plans for the Kootenay Valley and Kootenay Lake.

Anderson and Gilmour produced a map that functioned as a key coordination and planning instrument. It drew together timber maps, the location of permanent and portable sawmills, headquarters, ranger stations, trails, roads, telephone lines and a series of strategic lookouts. That year, 1200 miles of trails, 10 ranger cabins and 300 miles of telephone lines were put in place. But the true star of the fire protection system was the lookout tower. The new Mount Baker lookout, at 7100 feet above sea level (2164 metres), commanded a view over a million visible acres that included settlers and the smoke plumes from 26 sawmills and planing mills. Even before the telephone line construction gang had reached 4500 feet (1370 metres), they spotted a fire

Black Mountain Lookout, 1931. Lookouts allowed the fire watcher the ability to see a wide territory. NA-05885.

Checking weather recording instruments, Gonzales observatory, Victoria, 1956. NA-15842.

eight miles away that was put under control that night,[7] as speed and accurate information also allowed for efficient use of fire crews. In an effort to increase that efficiency, bids had been put in place to construct four 36-foot and two 54-foot launches for the coast, plus two 25-foot launches for the Arrow Lakes and Kootenay Lake.

In 1915 the Forest Branch founded another Dominion partnership, this time with F. Napier Denison, superintendent of the meteorological station in Victoria at Gonzales Height. Denison began supplying the Forest Branch with daily weather reports for all of British Columbia. Specialized reports began to be made for the times of highest fire danger, with the collection of temperature and precipitation information. Examples of district weather charts for Quesnel, Vernon and Vancouver's fire seasons appeared in the annual report for 1915, although humidity indexes had not yet appeared.

Mixed Use: Grazing and Range Management

Both the Forest Branch permanent staff and the survey parties were tasked with identifying and documenting existing and potential range and grazing lands. Primarily cattle- and horse-oriented, the interface between pasturage and timber is more porous than many would imagine. Initially under a combination of the *Forest Act* and the *Crown Pastures Act*, the use of lands by cattlemen provides them with assistance as grazing reduces the fine fuel load in the interface between timberlands and open meadows. The early estimates showed that while three million acres were available for pasturage, a further 10 million acres of timberlands were also capable of sustaining and sheltering cattle and horses. Properly managed for carrying capacity, grazing on Crown lands helped sustain the BC cattle industry and assisted small operators who could not afford to acquire the land needed for large operations. As an example of scale, the Chilcotin's Gang Ranch, founded in 1874, reached 16,000 square km of grassland.

Range management was undertaken early as one of the areas of the Forest Branch with high rural visibility. It is an aspect of Forest Service history not well understood by urbanites, but it remains to this day a core part of the organization's mandate. Range management also supported and, at times, conflicted with

Cattle on the Edwards Ranch in the North Thompson region. B-03926.

the agricultural operations of First Nations communities, particularly where it involved free-range herds of horses. Ranching operations provided trails, eyes on fire hazards, partners in dealing with invasive plants and a ready source for community aid in fire season. In an age of horseback survey parties, the ranchers, outfitters and pack trains were key parts of transportation, whether in taking supplies to cabins and lookouts, assisting in district surveys and trail construction or just passing advice. In addition, the Forest Branch found the traditional range associations, agricultural fairs and rodeos good mechanisms for distributing information and concerns about range and forest issues.

The Hand-logger

Much celebrated in British Columbia's logging culture, hand-loggers used axes or handsaws to fall trees. They used gravity, assisted by jacks and cables, to get their logs to water, where they assembled booms of logs to be towed to market. The hand-logger licence first appeared officially in the 1888 *Land Act* as a general licence. It allowed any person to cut timber on any Crown lands which were not under a timber licence for one year for a $10 fee. Originally, there were no restrictions on how the person logged. They paid the same royalty rate as a timber licence. In 1906 the law was changed to forbid a hand-logger from using steam machinery, such as the donkey engine. Two years later the fees increased to $25. The licence was restricted to the northern coastal region, and operations became subject to inspection. Only someone appearing on the registered voters list or who was First Nations could apply for a hand-logger licence.

By 1918 the Forest Branch was not very happy with its 210 hand-logging licensees. They were highly critical of the number of felled trees that were abandoned when it proved too difficult to move them to water. A Commission of Conservation report, using information from Chief Forester M.A. Grainger, charged hand-loggers with destroying 1000 miles (1609 km) of coastline to an average depth of five chains, cutting 500 million board feet but leaving 300 million board feet in waste and indirectly causing the loss of another 800 million board feet through fire and windfall. Grainger agreed the hand-logger had a place on steep slopes but was unable politically to end the licences: "the usefulness of the hand-logging system has passed … it should be discontinued, as inimical to the object of forest conservation." Small timber sales, which allowed steam machinery, would slowly replace the hand-loggers, but they allowed access to those who lacked capital or were unwilling to accept wage labour as contractors.[8]

A hand logger at work with his jack. D-05561.

Early Entomology Partnerships

A forest entomologist of today would be at home with the contents of the 1914 annual report which discussed insect damage to merchantable timber. The problem had been known since at least 1904, and complaints during the first two years of the Forest Branch brought the problem into focus. An arrangement was therefore entered into with Dr Gordon Hewitt, the Dominion entomologist, and cooperative projects began. The first studies were undertaken by Professor J.M. Swaine, a North American expert who spent the summer of 1914 studying destructive insect outbreaks and collecting insect samples in the Kootenays, Similkameen, Okanagan, lower coastal region and Vancouver Island. His report to the chief forester identified three problem insects: the Western Pine Beetle (*Dendroctonus brevicomis*), the Mountain Pine Beetle (*D. ponderosae*) and the Red Turpentine Beetle (*D. valens*). He noted that the Red Turpentine Beetle outbreaks tended to follow the other two. While he ranked the Mountain Pine Beetle as the worst, the Western Pine Beetle was almost as frequent and damaging. The worst outbreak Swaine studied was among what he called Bull-pine or Western Yellow-pine, now Ponderosa Pine, at Princeton. Allowing for soils and climate, he felt that species vulnerability was the deciding factor, not location.

Pine bark infested by pine beetles, 1930.
NA-05808.

> The trouble is evidenced by clumps of dead trees, "red-tops," which is now becoming numerous throughout the valley and upon the mountain-sides. When examined this summer the clumps of red-tops contained five or six to about thirty-five trees and will now be considerably larger. The beetles had left the dead trees and were to be found in immense numbers cutting their tunnels in the inner bark of the trunks of green, healthy trees surrounding the clumps of red-tops. Of the three species involved in this outbreak the western pine bark-beetle was apparently the most destructive.[9]

In the worst case, Swaine found 1500 to 2000 pairs of beetles on a 50-foot length of trunk, burrowing, tunnelling and laying eggs inside chambers. Defensive gum tubes and red boring dust from the beetles marked the infested trees. In the spring the beetles emerge through round holes and move on to fresh trees. It might take three years to fully kill a tree, but once the tunnelling was begun, the tree was literally opened up for larger boring insects to enter.

Swaine set out the earliest treatment advice: floating logs to drown the beetles, winter cuts and slab burning or, in the worst case, cutting, piling and

Bark beetle infestation of Western Yellow Pine 1923. NA-03897.

winter burning before the insects emerged again in the spring. He noted in his 1914 report that the Sugar and Mabel Lake Lodgepole Pine outbreak of Mountain Pine Beetle had been underway for eight years since the CPR first noticed it, and another related species of beetle was attacking the Engelmann Spruce nearby.

Swaine identified a wide range of lesser beetle insect problems, at locations such as Cowichan Lake and Campbell River, as well as the combination of beetle and spruce gall infestation that was threatening Vancouver's Stanley Park in 1914. In his report of that year, he also discussed ambrosia beetles, fungal infections of Western Redcedar and closed by noting that the Spruce Budworm, "which was very abundant in many places a few years ago, is now hardly to be noticed."[10] By 1914 forest entomology for the next century was clearly defined in terms of the major threats and a basic understanding of life cycles and potential industry responses.

The 1913 Recession and the Great War

Despite such a remarkable start, all was not well for British Columbia's forest industry. Inflation and the economic downturn of 1913, coupled with the over-reliance on a single prairie market which was based on immigration that had slowed, meant the full-capacity forest industry was extremely vulnerable. The 1913 downturn also meant aggressive market competition from American producers.

By 1914 struggling companies could not meet their stumpage payments or rents, and the emergency response was to provide relief under a new Royalty Act. The relief aspect of the act allowed for a delay of up to one year in payment of royalties, buying time for the producers and delaying the surrender of licences. The declaration of war in 1914 compounded the problems by bringing the industry to, as MacMillan noted, "a dead halt" due to the complete paralysis of the building market. Marine shipping also immediately became scarce.[11]

Logging at Summit Lake, 1912.
C-05269.

As the war developed, Britain's Imperial Munitions Board sought food, munitions and war supplies. Shipping losses to U-boats meant a shortage of ships for low-value items such as lumber. Many of British Columbia's underemployed loggers rallied around the Union Jack, and communities from Prince Rupert to Nelson assembled volunteers who went into the great numbered battalions that eventually became casualties of war along the western front.

Less well known is that specific forestry battalions were recruited to work in French and Scottish forests. Canadian loggers, foresters and Chinese labour battalions sent to Europe provided much of the wood used in constructing trenches, tunnels and bunkers, as well as building over 100 aerodromes. Every tree cut in Europe saved space for other munitions in the Atlantic convoys. For the Canadian Forestry Corps, experience overruled the usual medical rules about age and fitness. For instance, Corporal Bates from Ganges on Saltspring Island was 56. At least a dozen volunteers were accepted who were missing fingers, and one volunteer from Vancouver had already lost his left hand.

The Forest Branch began to lose staff as so many volunteered. From 1914 to 1918, staff dropped almost a third, from 167 to 114. But as the editor of

Martin Grainger and forester J. Latham setting out on an inspection trip from the Hazelton Forest Service Office, 1914. H-05547.

the *Forest Branch Newsletter* pointed out in the November 10, 1939, issue, the honour list on the wall at the chief forester's office shows 129 names, the equivalent of 76 per cent of staff, which reflects the coming and going during the World War I years.

Seasonal workers also became scarce. Therefore, logging operations, particularly in the interior, faced not only poor markets but their own staffing shortages, and they owed royalty and rent payments they could not meet. So Forest Branch revenues dropped by another 12 per cent, which meant fewer crews could be fielded for survey work or fire protection infrastructure. During the war years the Forest Branch had lost half its temporary and permanent workforce, dropping from 558 in 1913 to 278 by 1917. But the most noticeable impact of the war on the Forest Branch was in the field, following the resignations of over two-thirds of the assistant rangers. The 1917 year-end report was itself primarily statistical "owing to the heavy enlistment among staff and the pressure of current work (especially in connection with speeding-up aeroplane-spruce production)."[12]

Building Global Markets

The problem of markets brought the need for leadership on that issue to the door of the chief forester. The dual crises of economic collapse and war

Barges loading export timber onto the *SS Grahamland*, Victoria harbour, 1915. NA-03902.

pushed the BC Forest Branch directly into the need to find markets on behalf of the province and industry. In 1915 MacMillan was appointed by the Canadian government as a special trade commissioner. His tour of the British Commonwealth included markets in South Africa and Australia, and he successfully lobbied the Imperial Munitions Board for contracts for British Columbia. In London he and former Premier McBride were horrified to find US wholesalers were re-stamping BC lumber as American and selling it to the Imperial Munitions Board while initially sitting out the war. MacMillan had a war prize that had been seized in the Falklands, the merchant ship S.S. *Grahamland*, assigned to bringing BC lumber to market, but that ended on an ill-fated voyage when the sides of the Panama Canal collapsed. The scale of the Imperial Munitions Board's contracts were too large for any one BC company. They were successfully fulfilled with the aid of the Forest Branch coordinating local firms, including the specialized market for aircraft-grade Sitka Spruce (*Picea stichensis*) for the air war. MacMillan's activity on behalf of the Forest Branch was important in that it staunched the downward spiral of lumber prices and refocused the province's industry on global markets

The promotion of BC wood also came to the fore during this time. Internationally, it became clear that major markets were unaware of BC production, or they had misconceptions about the qualities of the wood. International education, such as the first BC wood exhibit at Yokohama and work in South Africa, became an important activity for the Forest Branch. To develop domestic markets, the first bulletin publications were made. A series of 10 booklets on how to use wood to build specific types of farm buildings was put together, using university expertise from agricultural experts in Saskatchewan and with a foreword by that province's Minister of Agriculture. A quarter of a million of these simple booklets were distributed free to farmers across the prairies.

MacMillan's work on behalf of BC opened his eyes to international markets and opportunities in private industry, using the connections he had developed. But when he returned to British Columbia, he found the war had decimated the Forest Branch staff. His best and brightest had, in many cases, left for the war, and few ever returned. On September 12, 1916, using a personal code based on the Bible, MacMillan sent a telegram to the Hon. W.R. Ross, the Minister of Lands, tendering his resignation and recommending Martin Grainger as his replacement.

Grainger not only had big shoes to fill, he had to contend with a shrinking staff and a new minister to work with, as a decade and a half of Conservative government gave way to the new Liberal government. The minister now responsible for the Forest Branch was the Hon. T.D. "Duff" Pattullo, who for most of his career (which would later include being premier) had a deep interest in resource development, including forestry, hydroelectricity and early oil and gas exploration.

As a chief forester, Grainger's biggest challenge during the war was not the slow and steady development of domestic markets in eastern Canada, the shortage of shipping for Pacific markets or how to rebuild staff levels or deal with the fire season. The biggest issue came late in 1917, and it involved a marginal species: Sitka Spruce.

Remembering Grainger

R.V. Stuart: *It would be hard for me or anyone else to do justice to Mr. Grainger. He was one of the most lovable characters that I ever met. He was a man of high intelligence – his academic accomplishments were very high, and he was a delightful person. He was a wizard at mathematics. But above all this, he was the man who instilled more "esprit de corps" into the Forest Service than anyone I ever knew.*

In 1920 I got the appointment of assistant forester. Grainger used me from time to time around the province – I was what you might call a "trouble shooter." He was a man who was 100% sincere. His position was very difficult, for, you see, he was the chief forester when the new government came in, and the old Conservative government had been in power for 15 or 16 years. People naturally expected that there would be a lot of changes. Grainger sold himself and his service to the new government. There was no disturbance whatsoever. That was quite an accomplishment. You and I, who know something about politics in the east, would know that. Grainger introduced a lot of innovations which were very constructive measures in the Forest Service. During his time we introduced the matter of competitive appointments for nearly all Forest Service positions. This put the Forest Service almost entirely out of political control. During Grainger's time the Royalty Act *had to be overhauled, but he left before that was completed. He brought the industry and the Forest Service close together. MacMillan hadn't done it because he wasn't there long enough. Now whether he would have done it later is another matter.*

You remember we always had a problem here about the export of surplus logs. Up to the time that Grainger took over there had been no medium for doing that. It was all a matter of someone having more political influence than the others, and they would come to Victoria and get a permit. Grainger devised a system to do with the export of logs. It was the beginning of the Advisory Committee and handled that problem with the keenest taste.

Dr C.D. Orchard: *It was a wonderful institution and was still in existence when I left in 1958.*

R.V. Stuart: *Now there was another thing Grainger did. The loggers and millmen were always arguing about scales. A lot of free scaling organizations were started. The Logger's Association started one, and then I became manager there. We had as many as six scales. The mills individually and through the association added another dozen. Well, Grainger solved that problem by introducing what he called the "joint scaling and grading" way. He put three millmen and three loggers on the board, and it was presided over by the district forester.... Grainger made no pretense of having any knowledge of technical forestry. He was sold on the proposition 100 per cent of getting sound forest management into the province as quickly as circumstances permitted.*[131]

Martin Allerdale Grainger

Born in London Nov 17, 1874, and raised in Australia, Martin Grainger studied mathematics at Cambridge for his degree. Seeking adventure, he and two friends came to British Columbia as part of the Klondike Gold Rush of 1897 and, short of funds, lived and hauled freight in the Stikine River area, learning the axe, horse and paddle and helping in a government rescue of stranded miners on the Liard and Dease Rivers.

In 1899, he sought new adventures and despite flat feet went overseas to serve as a Lieutenant in the Boer War. There he also wrote dispatches to the London Daily News criticizing the poor information and supplies that hampered local operations. In 1901 he returned to work with oxen, logging on Pender Island and meeting his future wife, Mabel Higgs. Two years later he went to work as a handlogger cutting and booming logs at Knight Inlet, later described in his fictionalized and semi-autobiographical novel *Woodsmen of the West* first published in 1908, dedicated "To my creditors affectionately." It is a classic of Canadian literature and has had several editions.

His work as secretary for the Fulton Royal Commission came in part due to talent, partly by political connections, having written speeches for Commissioner Alfred Flumerfelt, and rumours that a predecessor at the Commission was removed due to their fondness for drink. Chief Forester H.R. MacMillan credited Martin Grainger and American conservationist Overton Price with crafting the *Forest Act* of 1912 and creating the new

M.A. Grainger crossing the Lakelse River on a log jam, 1914. H-05554.

Forest Branch to monitor the forest industry. He introduced concepts like sustained-yield to the report. As Chief of Records he drew on his Boer War experience to see the need for good information on the ground for staff. Yet he preferred fieldwork and in 1914 set out on a tour of the Hazelton district, inspecting Forest Branch cabins. However a heavy push by foot to meet his wife at Lakelse Lake permanently injured his feet. When he could walk again, Grainger wore moccasins, even when he met King George V and Queen Mary in London in 1920.

In World War I Grainger became Acting Chief Forester while MacMillan was travelling and replaced him when he resigned, becoming the second Chief Forester in January 1916. Grainger changed the direction of the Forest Branch from marketing BC's lumber to managing timber, with a regulatory focus. He hired staff on merit, and avoided political appointments, to make the Forest Branch more independent. He launched public education programs and favoured research to reduce the lower-grade timber. While he viewed old-growth forests as a form of waste, he did argue for increased stumpage to pay for fire-fighting and regeneration. In July of 1920 he represented British Columbia at the Imperial Forestry Conference in London which examined supply issues across the British Commonwealth.

On his return, and increasingly frustrated with government and industry opposition to increased stumpage and regulation he resigned in 1920. In a subsequent speech he expressed disappointment at the lack of influence of the *Forest Act* over forest practices, and commented that the West should take a greater interest in its own future, especially considering that 'the world has never seen timber massacred on such a scale before. There is more knocked down and wasted than ever grows in most other countries.[14]

Yet in direct contrast to the speech in 1921 Grainger became director of the BC Timber Industries Council, and lobbied against raising stumpage fees, his own Forest Branch proposal. He left the Council in 1924 after its campaign against a minimum wage and an eight hour day, to work as an independent forester and broker for British timber investors but the Depression eroded that business.

In his private life he fell in love with the forested landscape around Princeton that he first saw on horseback with Ranger Bert Thomas. He rented and then built a cabin in the area. It was there he wrote the sketches which became his next book, *Riding the Skyline*. In 1929 a Depression highway project began, building the Hope-Princeton highway. Grainger began a campaign to have a park created in the region before land prices rose and bring people into the local economy and the landscape he loved. He also fought with Keremos sheep ranchers over range versus park use. His replacement as Chief Forester, Peter Caverhill, created as a compromise Three Brothers Mountain Game Reserve in October 1931.

M.A. Grainger's cabin on One Mile Creek, Princeton. C-07687.

Martin Grainger died on October 15, 1941, in Vancouver at the age of 66. Earlier that year, the Hope-Princeton Mountain Park that he had lobbied for was established, renamed Ernest C. Manning Park after the Chief Forester who died in a plane crash that year. A man of contradictions and complexity, most comfortable on a horse in a forest, his classic novel about work in the woods was rediscovered and reprinted in 1964.

Jenny Clayton

Spruce for Aeroplanes

Early aircraft were built of "doped" linen (strengthened with a dried flammable liquid) stretched over a wooden frame, brought into tension with struts, metal tubes and wires. The interior of the aircraft's body and wings were all wooden construction. At first, hardwoods such as ash and hickory were used. As designs were refined, the search was on for a wood with the right balance of light weight and great strength. The wood needed for mass airplane production had to be plentiful, with a clear grain, easily milled, resistant to rot and easily harvested. Just before the war, aircraft designers hit upon Sitka Spruce as an ideal wood for all of these reasons.

Aircraft were the eyes in the sky used to direct artillery and to create the maps for planning attacks in the dense, gridlocked trench war. They battled other aircraft and giant balloons, and they defended and bombed navies

Logging Sitka Spruce near Terrace. E-01790.

and cities. The French-made Nieuport Scout, the British Sopwith Camel and the Handley Page heavy bomber were just a few of the many types of aircraft flown by the young pilots of the Royal Flying Corps, which included young pilots from British Columbia.

As the air war over the trenches intensified, an ever-growing number of biplane aircraft and pilots were lost. Allied factories, primarily in Britain and France, were going around the clock producing aircraft and used aircraft-grade Clear Spruce from the western United States. Late in 1917, though, American spruce became in short supply for two reasons. The first was the US entry into the war and its own demands for aircraft production. The second was a series of longshore labour problems in American ports that threatened to choke off exports. Spruce production was of such importance to the US war effort that it became a completely military operation, with 19,000 soldiers working from five dozen military logging camps during the war.

R.V. Stuart on the Vancouver Office:
I was transferred into Vancouver in September, 1916. In those days the job of district forester didn't rate an assistant. I was there to help the D.F. in a position which was then called the "lower ranger" and had as a district all the territory of the Vancouver forest district east of Point Atkinson. MacMillan came into the F.S. offices in March or April of 1917 (after he left the F.S.). At that time I was in a little office alongside of Van Dusen. Mac came in to see Van, and as Van had someone in seeing him, MacMillan put his hat on my desk and said that he was looking for a job. Well, the next thing that happened was that they found Sitka Spruce was a wonderful wood for making aeroplanes, and the word came from England that they wanted all the spruce that they could get. They applied to Victoria for someone to take charge of organizing the logging and getting it shipped over. Mac was at loose ends, and Grainger recommended him. After Mac had been on the job a month, he took Van Dusen.[15]

For the Imperial Munitions Board, the solution to aircraft production for the war grew in the forests of coastal British Columbia. Large stands of Sitka spruce were to be found on Vancouver Island and along the coast, but especially in the Queen Charlotte Islands (Haida Gwaii). Top-grade aircraft spruce came from trees 16 feet (five metres) in diameter and 328 feet tall (100 metres), often 400 years old. Brought to water, they could be rafted down to Vancouver for milling into aircraft-grade cants. Yet only about five per cent of a tree was finally used as No. 1 aircraft-grade lumber, though uses were found for the rest.

The Imperial Board sent a representative to British Columbia to discuss the supply shortage of what was now a strategic material for the air war and then established a Dominion Branch of Aeronautical Supplies. Negotiations for cutting rights and permits went slowly, far too slowly. The provincial cabinet stepped in, issuing an unprecedented emergency order-in-council authorizing the Minister of Lands (and, in turn, the Forest Branch) to authorize the immediate logging of all Sitka Spruce on vacant Crown lands for the Imperial Munitions Board's needs – much like the ancient Naval Stores reservation used for navies in the 18th century. This order was followed shortly after by the Spruce-cutting Act, which provided compensation at $6 per 1000-foot board measure for No.1 grade logs and $2.50 for No. 2 grade logs. The rate was high, designed to cover any related waste due to species-selective logging.

A literal army of machines and loggers were moved up from the southern BC coast to invade the Queen Charlottes. In January 1918 a mere 116,000 board feet of spruce were cut. Eight months later, the cut for the last month of the war reached 6.8 million board feet. In total, BC cut and shipped the equivalent of the entire western US cut, some 26 million board feet of Sitka Spruce and nine million board feet of Douglas-fir. Just as quickly as this demand started, the war's end brought an abrupt shutdown of spruce production. It was wasteful and destructive from a resource standpoint but also a major factor in winning the war of attrition at the front.

The Queen Charlotte Islands Aeroplane Spruce Company logging Sitka Spruce, about 1944. D-04660.

Chapter 3
The Forest Branch, 1919–1945

Emerging from One War

After the armistice in November 1918, the economic momentum the war had given to British Columbia's forest economy continued. The British government's timber controller visited Vancouver, and an order for 70 million board feet was placed for postwar rebuilding. But it was no longer supplied or coordinated by the Forest Branch, rather by Vancouver's newly established Timber Export Board of British Columbia. The timber controller's visit was followed by visits from other British buyers and a trip to view sawmills and logging operations around Vancouver for the Prince of Wales in 1919.

Prince George Rangers Meeting at Aleza Lake, 1928. Left to right: (front row) Sansom, Langmuir, Lowry, Dr Lashley (Giscombe), Kerkhoff; (middle row) Norman, Faherty, Mathieson, Wilson, Orchard; (back row) Forbes, Hunter, Russell. NA-04104.

Canadian domestic markets also began to pick up as the housing industry restarted. Over two million people saw the exhibit about BC wood products at the Toronto Exhibition. The Western Redcedar shingle and shake producers found their business began improving on the prairies, and the forestry industry generally saw a return to the pre-1913 optimism.

Also returning were soldiers, which was a relief to the Forest Branch. Veteran retraining programs included forestry, and forestry work appealed to those who had served in the Canadian Forestry Corps, those who sought a wilderness lifestyle after trench warfare or who were unwilling to return to cities and factories. Their war experience made work in the Forest Branch a comfortable fit.

The returning soldiers sat the new ranger and assistant ranger exams. These exams, designed to ensure certain standards in the service, were balanced with the need for new staff having woods experience as well as surveying and firefighting skills. Chief Forester Martin Grainger reported in 1919 that practically "every vacancy in these positions" was filled by a returning soldier.[1] This renewed and reinforced the military model of the Forest Branch as an organization. Work on foot or on horseback in the quiet woods must have been a dramatic and welcome change from the mud- and water-filled trenches and deafening explosions.

The postwar era also saw the launching of public education campaigns. Fire restriction notices were combined with cartoons placed in each district's newspapers, their appearance planned out using estimates of risk based on weather and fire histories. By 1921 the Forest Branch, working with the shingle manufacturers, entered the film industry, producing short features showcasing the BC industry and sending them to London in order to promote markets and investment. The earlier farm bulletin program aimed at settlers and prairie buyers continued, but the new bulletins emphasized the strength of BC woods and were aimed at architects and builders. Samples were sent to educational institutions, and advertisements were taken in trade journals.

Range Management and Grazing

1919 was also an important year for an area that many urban British Columbians are unaware of as part of the Forest Branch's mandate and history: range management. The provincial government declared a new *Grazing Act* in March, and this new act reflected community consultation. The previous winter, Forest Branch staff attended stockmen's meetings, conducted range surveys and, in conjunction with the ranchers, came up with guidelines and acceptable forms that were approved by representative committees before the legislation

B.M. McRae's herd of cattle, Topley. C-07867.

was drafted. Yet the system made assumptions about education and literacy appropriate for a city but not for a small rancher who left school to work early, and many a staff person found themselves sitting down at the kitchen table on a small ranch helping the family fill out their application. Seven grazing districts were established, with the largest stock regions being Kamloops, Nicola/Princeton and Okanagan/Similkameen, with 45,000 cattle and horses on Crown lands. (Interestingly, a cow and calf count as one animal unit when doing range inventories.) The Forest Branch also worked to help organize ranchers, where possible, into one of 17 stock associations, with the idea of associating them closely with specific range management areas.

One big shift in the new act was the requirement of public notice over application and a standard application closing date, July 1. It allowed those with long-term usage sufficient notice to come forward with their counterclaims. The previous unregulated approach, coupled with war shortages of staff, had created overgrazing issues and conflict among ranchers. The service's goal was more sophisticated management.

As the newly created Commissioner of Grazing, Thomas P. Mackenzie noted that, aside from a few large ranches, the ranching industry in British Columbia was primarily a patchwork of small family operations – consisting of 25–50 head, some having as many as 250 – and that he expected it to remain as such. Grazing in BC was often limited by landscape. Grasslands tended to follow rivers in narrow, V-shaped valleys with steep slopes where the treeline and grasslands tended to make a slow transition into each other. And so BC's landscape also did not easily favour much in the way of cattle drives. The result was an industry, then and still today, based around small, compact, family operations.

Range restoration was also a big issue. The need for beef for the war effort had a cost in destructive overgrazing. Areas were now assigned for spring and fall stock movements, and by 1920 management plans in a few areas began to look at protection for bunchgrass and blue-grasses through brief seasonal restrictions during the growing season. Another area of friction was the issuance of cutting licences for wild hay in the mountain meadows, seen as a commons. Forest Branch grazing and range staff were also not happy with First Nations' traditional open range practices, where horses were generally left to roam in herds until needed. Other pressures on the range included wild bulls and scrub cows, and the so-called *"Good Bull" Act* of 1920 was an attempt to identify these for removal, both to produce more profitable beef cattle (Mackenzie favoured Herefords for stock improvement) and also to remove pressure on range.

The Grazing Office established its own research herbarium in 1920 with examples of 116 grasses, 34 sedges (grass-like) and 294 herbaceous plants,

Cattle roundup near Kamloops, about 1900. H-05836.

Bunch grass in the Salmon River area, 1914. NA-04162.

all gathered with the assistance of Dominion botanists and Professor John Davidson of the University of British Columbia. The purpose was to understand the utility of plants and their individual cycles as they related to grazing. Invasive plants were not yet an issue.

There were many points of contact with Dominion staff and range managers in other jurisdictions as well. For example, the grasshopper was cattle's tiny but furious competitor for fodder. The grazing commissioner and his staff worked with the Dominion entomologist and were aware of successful grasshopper control programs in central Oregon and South Africa. A 1920 experiment at Riske Creek, BC, by Dominion entomologist E.R. Buckell found bunchgrass to be relatively free of grasshoppers, but 35 species of grasshoppers grazed among the blue-grass, with only two types responding to poison bait.[2]

Provincial-Dominion cooperation on range issues occurred at other points, too, where ranchers sought range access both inside and outside the railway belt. This often brought Dominion and provincial counterparts together to consult and plan. Through the 1930s, Mackenzie's aggressive use of regulations brought him the ire of many stock associations. His experiments on range at Tranquille, near Kamloops, had mixed results, and with the economic depression, his team was dispersed and the operations at Tranquille shut briefly until federal funds allowed it to restart.

Lines by a Forest Officer, Newly Introduced to a Grazing District.[3]

"Pray tell me Sir," the Maiden cried, "how earnest thee thy keep?
A Forest Ranger yclept thou art, but rangeth naught but sheep."
"Alas, 'tis true," the Ranger said, "Mine head in sorrow bows,
No joy exists, the world is dark; I'm cursed with sheep and cows.
"My days are filled with duties dark, my nights with anguish deep,
For I, who once cruised stately Spruce, now cruise the noisome sheep.
"For years I trod the forest aisles where lofty pinetop sweeps
The clear and balsam-scented blue; the dark green mountain steeps.
"Where oft I've watched the clownish cub; the deer their vigil keep,
While slaking thirst in limpid stream where finny elfin leap.
"Where stilt legged Moose and antlered Elk came bounding into view
Now steals the odoriferous Ram in search of virgin Ewe.
"The sylvan glades and cooling shades are gone as last night's sleep. Alas!
So has the peaceful quiet – the slopes are Baa-d with sheep.
"And though, sweet Maid, I range the hills, no more their praise I swell;
Though only He can make a tree His man can still raise Hell.
"The timber lies entangled now in scarred devasted heap,
The green of mountain meadows fades to grey with bleating sheep.
"Midst contemplative cows – wend and wary must I tread,
No mossy log to rest my bones; the ground is spread with spread.
"The trees and alpine flowers fade and die as Heaven weep,
For how can green things grow, fair Maid, in air perfumed by sheep?
"Mock not my woes; the mournful tale of war 'tween sheep and cow
Has driven better men than I to squirrel feed I trow.
"No jugs of wine, nor books of verse, nor e'en thy bright blue eyes,
Can ever turn this wilderness again to Paradise.
"Drift fence, mudholes, are now my Lot. Away the foul things sweep;
Lookout! I'm mutton: Ha, ha, Baa!!!
To Essondale . . . . . . I'm SHEEP!!!!

Building a Forest Atlas: Intensive Surveys

The return of staffing levels after the end of the war meant detailed forest reconnaissance work could begin again in preparation for topographic mapping. In 1921 there were six intensive surveys, which meant at least five per cent of the province's trees were measured, and two areas received a preliminary estimate based on a faster and wider timber cruise.

One of these surveys was related to a proposed sulphite paper mill for Prince George and covered the valleys of the Fraser and Bowron lakes. This survey used 100-foot contours and was the first use of two new tools: the topographic Abney Level and two-chain topographic tape. The survey covered 22,000 acres (8903 hectares) of timber and identified possible slopes for railway logging access.

In the Nass region a preliminary 1914 survey was followed up with a more intensive timber cruise, providing a rough species breakdown (50 percent hemlock, 32 per cent spruce, 13 per cent true fir), but bad weather called a halt. The North Thompson River survey, previously halted by heavy early snow, was completed in 1921 and identified the area as suitable for summer logging and river drives. Initial work also began on burned-over lands between Clearwater and Myrtle Lake, providing insight into their potential for natural regeneration and commercial species mix, but this survey too came to a stop with an early

The Babine Forest Survey took several seasons. This 1927 survey was conducted by K.C. McCannel. NA-05645.

winter. Yet another crew mapped the forest-muskeg interface in the Queen Charlottes. These surveys were typical of the summer fieldwork needed to support the winter mapping work back at headquarters in Victoria. Conditions were rough, mixing transportation by boat, canoe and horses plus work on foot. The crews worked out of tents and temporary forest camps, amidst mosquitoes, horse flies, bears and threats of fire.

Surveys also produced data that was increasingly important to the Forest Branch research program. The linkage of measurements, species and location (slope or lakeside) meant staff could begin building a statistical basis for considering species, age, growth and local site information. Volume tables, sample plots and regeneration estimates became part of the language, partly for long-term forest management and also to answer questions around which area was best suited to pulp mill licences. This volume of data, some 175 tabletop sized maps, is what is known as the forest atlas.

A Quiet Revolution: Information Services

The quietest revolution underway in BC in 1937 was the introduction of punch cards for forest inventory data being implemented by the headquarters staff of the Survey division. "Cards are being prepared for the Electric Accounting Machine to establish a punch-card tabulating system for recording and sorting all the details of the forest inventory." The Hollerith punch-card system (invented to tabulate the US census of 1890) had, through a series of mergers, by 1924 evolved into the International Business Machines Corporation (IBM), which relied on punch-cards and tabulating adding machines before the invention of the computer. The Hollerith-based punch-card machines allowed for rapid and accurate tabulations of the data from the forest atlas. The machinery used by the Forest Branch was developed in cooperation with the Bureau of Economics and Statistics of the Department of Trade and Industries. The first major project was to review the Vancouver forest district atlas and inventory. Numbers were converted onto punch-cards, bringing 125 of the 183 watersheds into the data system. It was the first step in what we call the computer revolution for forestry.[4]

The Forest Branch's library began informally, in scattered offices. One could say it originated in the 1932 centralization of 988 text and reference books, bulletins and pamphlets collected from various offices and indexed to establish a small, efficient Forest Branch reference library. By 1937 there were 2673 forestry publications carefully indexed. As an information service, the Forest Branch library collected recent and older publications, including technical papers from laboratories, forest services and research stations around the world. The branch was soon subscribing to 19 trade journals and periodicals, and a further 35 periodical subscriptions were received as gifts. All articles considered valuable to branch staff were catalogued; the card catalogue grew to 12,000 entries, while a further 10,000 cards of forestry material could be found elsewhere, such as in the provincial library.

Chief Forester Peter Caverhill and the Great Depression

In 1910 the University of New Brunswick's new School of Forestry graduated its first class. The four foresters in that small group included Peter Zemro Caverhill, who joined the BC Forest Branch in July 1912, serving as district forester at Kamloops. In 1916 he worked briefly as chief forester in New Brunswick but returned to BC in 1917 to take the position of district forester in Vancouver. Caverhill worked his way around the Forest Branch before once again returning to New Brunswick in 1917, this time to complete a Master's degree at UNB.

A highly trained technical forester, Caverhill would expand BC's Forest Branch into new areas, as a forest science organization. He established the core research programs and created new research stations. He also completed the province's forest inventory atlas system, began its silviculture program and built its nurseries. It was under Caverhill that new technologies were introduced, such as radios, aerial photography and, for the records and research division, a central library and card-based sorting system of forest data. In the timber sales office 1923 marked the appearance of the "ditto" machine, which used different colour wax backing sheets to produce typed and drawn copies of documents – a purple, typed description and a red timber sales map drawn to show cut area. Its older competitor, the mimeograph stencil, could only produce one colour.

Views of Caverhill

Perhaps one of the most famous memos from Victoria to reach a ranger in the field was issued in the Depression by Chief Forester Peter Caverhill – who, if he is remembered at all, is remembered as the bureaucrat who once instructed field staff to use a pencil down to the nub before asking for a new one. He was a bean-counter in the view of many rangers. But the few who knew him well, like R.V. Stuart, spoke otherwise when he and C.D. Orchard were interviewed:

Now when you come to Caverhill, there was a man who, to my mind, we don't treat with enough respect for what he did accomplish. He was the man who really carried the Forest Service through the troubled years of the Depression. It broke his health and killed him. Anyone who wasn't around here in 1931–33 could have no conception of the tough times he went through trying to keep the Forest Service going. Dick [Orchard], you were there, so you know how it was. You can't compare Caverhill with Grainger or MacMillan or any of his successors. He was a man of his own type – very reserved. But I never met a finer man in my life, bar none. He was a top administrator. If he had any fault, it was he was too much concerned with the influence of pubic expediency. He was holding a high office in the Tolmie government, which was at its wit's end as to what to do for money. No one before, or since, has had such a struggle. But I had great respect for him. Caverhill died in 1935, and I was one of the pall bearers. Well, I got to know him … in a sense I grew up with him here. He was a bit older than I was. I look back to the time, late 1914 or '15, when we would go out on woodland trips and camp out overnight and so on. His outstanding characteristic was reserve. He was a shy fellow. He could get up and give a nice graceful address, and he had a sense of humour. But you couldn't have made a public speaker out of him as you did Manning, or with the inherent ability of MacMillan.[5]

The *P.Z. Caverhill* in Vancouver harbour, 1940. For a time, the BC Forest named the vessels in their fleet after prominent members of the service. NA-07076.

Caverhill also sought partnerships with the Dominion of Canada. One of the most successful was in taking advantage of its expertise in aviation, using the war surplus aircraft Britain had given to Canada. From 1920 to 1924 Canada had a civilian Dominion Air Board which oversaw the postwar civilian Canadian Air Force. The staff of veteran pilots undertook forestry, border and survey work before becoming part of the National Defence Department and being renamed the Royal Canadian Air Force in 1924. In February 1920 the Dominion Air Board established the Vancouver Air Station at Jericho Beach. The base cooperated with the Forest Branch and the summer of 1921 saw the first fire detection air patrols around the dense concentration of logging operations ranging up to Thurston Bay and eastern Vancouver Island, as well as the islands and mountains favoured by summer campers and hikers. Fifty-six bi-weekly flights went up, usually of three hours, on every second Wednesday and Sunday. In addition to providing a birds-eye view to the staffer who had a report of all active fires, flights also examined blow-downs and assessed the damage from a severe storm on January 29, 1921. While the initial impact may have been minimal, these flights established belief in aircraft as a tool for forest management in British Columbia.

By 1934 Chief Forester Caverhill reported that when a forest engineer crew conducting reconnaissance of the Nimpkish Lakes and forests on Vancouver

Curtis HL 2 at Alert Bay in 1920, prior to leaving on a survey trip. NA-07649.

Island was dropped off by plane, along with their canoe and two-month supply caches, they were put in place at four locations along the lakes in a few hours – work that would have previously taken a crew of packers and canoe-men two weeks or more. He also reported the completion of a forest map of the area, done with comparatively little fieldwork by plotting information from vertical air photographs taken for the photo-topographically controlled aerial survey.

Two years later, an early aerial survey project was done in conjunction with the E&N Railway land-grant. The forest surveys division organized its first large aerial survey project the next year, in 1937. An electronically operated British Eagle Three camera with a five-inch, wide-angle Ross lens recorded on a five- by five-inch negative, which allowed it to cover 3.6 square miles (5.8 km) on the ground from a height of 10,000 feet (3048 metres). Stereoscopic plotting based on aerial photographs completed a survey of Haida Gwaii, and the new images were combined with the older ones taken in 1933 by the Royal Canadian Air Force. Aerial photography was also used along the Skeena River between Kwinitsa and Terrace to calculate the volume of cottonwood forest in the region. Forest Branch headquarters staff watched this particular survey with great interest because the terrain along the river made such a survey very difficult and slow for ground crews.

The next year, the Forest Branch chartered a faster aircraft and had the benefit of bright, clear weather. Surveys were conducted of pulpwood areas at Prince Rupert, and mapping was done for Campbell River and the Okanagan watershed. Oblique photography was also tried for a grazing reconnaissance in the Kamloops district. As well, the service began putting its own stereoscope into use for photographic analysis, and in 1936 work began on panoramic documentation of the landscape from lookout stations. This was used as a method of locating and reporting forest fires. The project began in the Kamloops and Nelson forest districts and consisted of providing spotters and headquarters with matching maps and photographs so they could quickly and accurately determine the location of reported fires. Unfortunately, the program faltered due to the lack of resources during World War II but continued after the war.

Caverhill also oversaw the return to provincial control of the 14.4 million acres (5.8 million hectares) of federally administered lands – the railway belt, a corridor 40 miles (64 km) wide around the route of Canada's transcontinental railway. Since 1899 the federal government's surveyors and foresters in the railway belt had administered forests, timber berths and coordinated fire control with their provincial counterparts. But in 1930 the federal government returned those alienated western lands and their files to the provinces, and any willing staff was transferred to the BC Forest Branch. The federal headquarters had

Horse logging near Kamloops, 1941.
NA-07713.

been at the railway depot of Kamloops, and the Forest Branch took on that administration smoothly, which reflected good relations with their Dominion of Canada counterparts.

The Young Rangers of Palling

"These are our forests to defend, against the fires of thoughtless men"
– motto of the Young Rangers Band

In rural BC the ranger is often a pivotal person in the community's social life. In 1928 near Burns Lake, Ranger W.C. Saunders got word of a fire near a valuable stand of timber. A boy and a young girl each brought him the news. He rushed to the fire at Palling, west of Burns Lake, on the Canadian National Railway line and found the crew had the fire under control. Aside from two adults, the entire fire crew consisted of children and teenagers, boys and girls. They led a fairly isolated life on the homesteads. The adults refused pay, but Saunders found some money for the children for watching to make sure the fire stayed out. It set him thinking about what he had seen.

The result was the Palling Young Rangers, a secret forest protection society for 12-year-old girls and boys. It was run by the young rangers themselves with inner rangers and an outer forest guard. They had passwords, secrets and even a blindfold initiation involving walking barefoot on uncooked beans.

As well as performing fire watch patrols, the Young Rangers performed a mix of forest conservation and social activities, hiking, reforestation, picnics, summer camps and educational tours to lookouts. Saunders remained their proud supporter and advisor, assisted by Mrs Wilson of Burns Lake, and other rangers began to notice his region's excellent relationship with rural families. His area had a low number of fires because the Young Rangers also policed and critiqued unsafe family logging practices. The Young Ranger idea spread, and other lodges were established with the assistance of rangers at Smithers, Burns Lake, Vanderhoof and McBride. By the mid 1930s there were 300–400 members.

By 1937 the Young Rangers were clearing and building their own cabin on a one-acre island outside Burns Lake. That winter, they cut their own logs. In the spring local operators helped move them to water. They landed, peeled and prepared them for building. The island had been gifted by the Minister of Lands, Canon Hinchliffe, who on a visit had been delighted to have been kidnapped and initiated into the Young Rangers. In May 1937 Wells Gray became a member, and on his advice, on August 14, 1937, Lord Tweedsmuir, Governor General of Canada, and Lady Tweedsmuir visited Burns Lake. Both were inducted as Young Rangers and, taking the oath, raised their arms like tree branches, promising to protect forests from fire.[6]

First Young Rangers Band with their founder, W.C. Saunders, at Palling, BC. B-02175.

A System for Forest Research

It was Caverhill who instituted forest research in a systematic manner. The first researcher, James Alexander, was hired in 1921 and investigated timber utilization, regeneration growth, fire protection and timber volume. The results of the natural regeneration studies led to the decision to establish the Aleza Lake Experiment Station (1924), which examined sustained yield in spruce

Aleza Lake Experimental Station, 1930. NA-04037.

Camp personnel at the Aleza Lake Experimental Station, 1930. NA-04012.

and true fir forests, and the Cowichan Lake Experiment Station (1929), which established BC's first thinning plots. By 1930 the Forest Branch had Canada's largest research program, with two notable examples.

One was the charismatic Percy Barr, who had been a pilot in World War I. He was a UBC forest engineering graduate who joined the Forest Branch in 1924. Barr and his staff built the Aleza Lake Experiment Station, and from its data he completed and defended his Yale PhD on the ecology of spruce forests and spruce silviculture. Barr ran the research division but, with the Depression, saw budget cutbacks thin activities and staff dramatically. He himself left in the fall of 1932 to teach forestry at the University of California, Berkeley, where he ran the Blodgett Forest Research Station for two decades. Highly decorated in World War II, Barr returned to Berkeley and in 1953 taught the first course on industrial forestry in the United States.

Stig Schenström had worked on thinning projects from 1922 to 1928 in Sweden and graduated from the Royal College of Forestry in1927. Hearing positive things from a relative in Victoria, he joined the Research Division and set about creating the province's first and oldest thinning plots at the Cowichan Lake Experiment Station. The stands are unique because he carefully mapped

Stig Schenström, B.G. Griffith, G.H. Finlayson and J.D. Curtis at the Cowichan Lake Experimental Station, 1931. NA-04151.

out not just the trees but the forest debris in the stands, including fallen logs, allowing for historical measurement and calculation of Douglas-fir decay rates. Encouraged, he asked his fiancé to join him from Sweden. They married in Victoria in December 1930. Being neither a Canadian nor a British citizen, Schenström was the first employee laid off when money became tight. Finding little work in Canada or back in Sweden, he took a job tending plantations in Argentina, never knowing how reverently forestry professionals would view his early contribution, which is still studied today.

By 1936 only two researchers remained: Eric Garman and F.S. (Finn) McKinnon, a future chief forester. It was not until 1937 that Manning began rebuilding of the research division. Angus MacBean was hired from Ontario's forestry research branch, as well as George Allen from UBC. This division was housed in an office building at the corner of Michigan and Government streets in Victoria.[7] Forest service silviculture began in Victoria, on Shelbourne Street near McRae Avenue. In 1926 10 four- by 12-foot seed beds were put in place and planted with Sitka Spruce, Douglas-fir, Monterey Pine, Western Redcedar and some Japanese species. The Research Division, created in 1927, began as a research nursery supervised by Arthur Pickford, a specialist in Douglas-fir natural regeneration. It operated until its stock was moved to Green Timbers in Surrey in 1933.

On March 15, 1930, Chief Forester Caverhill opened a production nursery at Green Timbers. It was not without initial controversy. Local naturalists fought in vain to save the last stand of 200-year-old trees that were cut by King and

Buildings at the Cowichan Lake Experimental Station, 1931. NA-04257.

Stig Schenström, Percy Barr, E.C. Manning, Edward Prowd and Henry Cathcart at the Lake Cowichan Experimental Station, 1929. NA-04158.

Farris Lumber before they turned the land over to the Forest Branch for Green Timbers. Over the years the Green Timbers site grew to include an arboretum and experimental stands of Garry oak and Red Pine. It was also the site of the forestry training school for government staff. Today Green Timbers includes Douglas College and, by popular vote in referenda, a much loved major urban forest.

Experimental Plot 59, Shelbourne Nursery, Victoria, 1931, NA-03595.

Green Timbers nursery sign, 1940. NA-06967.

An important component for reforestation and research was seed extraction. The first attempts were initiated in the 1920s in a mix of private and federal facilities to supply Sitka Spruce and Douglas-fir seed. The Dominion plant operated in New Westminster from 1922 until, with the railway belt transfer, it closed in 1930. Internationally, it provided Sitka Spruce and Douglas-fir seed to Britain for replanting forests depleted during the war, and sent Ponderosa

Trees packed for shipment for planting on Thurlow Island, 1932. NA-04941.

Pine seed to New Zealand. Domestically, they supplied Dominion needs in the railway belt. Private operators also provided seed to BC's Forest Branch, such as Percy Ruth of Salmon Arm, who began his seed extraction operations in 1927.

Other early nurseries were also established around this time. For example, the Campbell River Nursery (1939) was a bare-root nursery created to assist replanting after the massive Bloedel fire at Campbell River which burned for close to 30 days and destroyed approximately 30,000 hectares of forested land (see page 106). In Duncan in 1943 work began on preparing a nursery between Chesterfield and Lakes Road, aimed at seed extraction. It came online in 1956, before the Koksilah (1967) location south of Duncan. Many other nurseries around the province would follow (see page 234 for a complete list).

The Death of Caverhill

As chief forester, Caverhill had dealt with the tremendous economic downturn resulting from the Great Depression of 1929. He watched his budgets shrink and his younger staff leave. He had to decide, virtually year to year, how to reorganize and what minimal tasks could still be supported, while still keeping one eye on markets and the other on the fire index and weather. Yet he never lost his sense of direction for the service. In reading his reports, you see a determination each year for a continuous, incremental introduction of new improvements and support for long-term research on forestry problems. Caverhill made it clear that he understood the Forest Branch's responsibility and held strong beliefs about its future despite any adversity. It was a tumultuous and inspiring 15 years for a remarkable chief forester, and it would end tragically.

On his return home from the 1935 Empire Forestry Conference in South Africa, a market of great interest to BC, Caverhill died unexpectedly of a massive heart attack a few weeks before Christmas. He had been ill in 1931 but returned fit for work. On the Friday, he went home with a cold, and on Sunday his heart stopped. Shocked, his staff gathered at his funeral to hear the acting premier, Wells Gray, praise him as a highly skilled forester but also "a friend and counselor."[8] For 15 years Caverhill guided the Forest Branch through good and extremely bad, but he never lost sight of the service's long-term goals. His sudden, unexpected death in December 1935 shocked the public, the government and the Forest Branch.

The Young Men's Forestry Training Plan (YMFTP)

Research centers like at Cowichan Lake might have disappeared if not for another aspect of the Great Depression: Dominion-run relief camps. They became scenes of discontent and strife, and a scandal emerged in 1932 when a cabinet minister's relative was seen to benefit from rent for the site of a relief camp. Concern that adult political unrest and protests of 1935 might spread to unemployed youth resulted in a new program to keep youth from being recruited into the politics of the relief camps. When Hugh Savage, the MLA for Duncan, stood up in the legislature, his call for a better system of employment for unemployed youth was well received. A budget of $90,000 was allocated in May 1935 to have the Forest Branch undertake an experimental training program to provide forestry employment for young men. Unemployment, ironically, saved and then expanded the almost dormant Cowichan Lake Experimental Station.

The goal of the youth program was not just to have young people do necessary work for the Forest Branch but was also, as the 1937 annual report emphasized, to provide "a valuable means of developing character, initiative, and self-reliance in the young men enrolled."[9] Activities included recreation; job training; field trips to sawmills and logging camps; forestry lectures by branch staff and visiting academic and Dominion scientists; first aid certification; log scaling courses; building and lookout construction; and trail, telephone line and firebreak construction and maintenance. Well received, the program was renewed for the following year. The Forest Branch model was successful, and the Dominion government agreed, after the second year, to pay 40 per cent of

Cowichan Lake Experimental Station, 1930. NA-04230.

University of British Columbia students transplanting at Green Timbers Nursery, 1931. NA-04803.

YMFTP's cost. The program continued to increase in size each year, with 860 men working in Forest Branch–run camps by 1939.

Hiring for YMFTP was done by a three-person board representing the Dominion, the Labour Branch and the Forest Branch. To be eligible, an applicant had to be between 18 and 25 years of age, have had some high school education and have been resident in British Columbia for five years. Successful applicants were paid \$1.75 per day, with \$.75 deducted for board, lodging and transportation. On the lower coast they went to one of three clearing stations. In 1939 those were the University of British Columbia demonstration forest, the Cowichan Lake Forest Experiment Station and the Quinsam Forest Nursery. Training for those hired in the northern interior was at Aleza Lake Forest Experiment Station. During the first two weeks, students attended lectures; watched forestry-related motion pictures; learned elementary forestry; were engaged in repairing structures, trails and paths; and learned to handle equipment for dealing with things like bringing down snags. After completing a probationary period, students were sent across the province to work in different locations, sometimes as assistants to forest rangers. Typical duties for YMFTP members included the construction and maintenance of pack horse trails, laying telephone lines, and constructing cabins, fire lookouts and bridges. They also removed hazards from strategic points and opened up areas previously only accessible by pack horse. Ranger Floyd Nelson was one of these students who went on to a career with the Forest Service. Two men who went through the

Trail construction, 1930. NA-04094.

YMFTP also worked as game wardens, and in 1939 five went to work as park attendants at Forest Branch–run parks. Due to the Second World War, only 17- and 18-year-old youths were eligible for the YMFTP training. The program's final year of operation was 1940.

But this was not the only program for the unemployed run by the Forest Branch during the Depression. The Forest Development Project provided work and housing at 21 camps in the Vancouver forest district for some 1400 single, homeless men. In 1937 Chief Forester Ernest C. Manning described it as "the first program in Canada designed primarily for forest development, protection and conservation, and reforestation as a means of alleviating unemployment,"[10] and the success of the Forest Branch camps brought interest from the Dominion in reforming their own troubled unemployment relief camp system.

The Forest Development Project relief program was aimed primarily at older unemployed men and transients. They worked for 30 cents an hour, based around a 44-hour work week. With federal funding, projects were undertaken in park and forest recreation. To improve the provincial forest reserves, they cut or rebuilt forest protection road access and removed hazards. These men did very valuable work, unseen by the public, on the watersheds for Greater Vancouver, Victoria and Nanaimo, building a system of roads, trails and fire guards to

Trail construction pack train at Meziaden Lake, 1930. NA-04030.

protect the watersheds. These projects provided work for 2809 men in the first year alone and double that the next year.

For many younger men, Forest Branch training, whether delivered through the National Forestry Programme or the smaller Forestry Training Plan brought them into direct contact with a possible career in forestry, as well as team spirit, healthy exercise and self-confidence. Through challenges like the assistant rangers exams, the training drew young men into forest work or gave them a direction when they returned to university studies. All in all, the Forest Branch could be proud of its role in helping the youth of British Columbia develop confidence and competency amid the harsh economic circumstances of the Depression years.

Ernest Manning's View of the Forest

Ernest Callaway Manning (1890–1941) became chief forester after the death of Peter Caverhill in December 1935. Manning was born in Selwyn, Ontario. His father was a farmer, and his mother was so determined to provide her three sons with an education they could not receive in Selwyn that she moved with them to Toronto and operated a boarding house there. Following his graduation with a Bachelor of Science in Forestry from the University of Toronto in 1912, Manning worked for the Canadian Pacific Railway office in Calgary, cruising for timber and scaling in the CPR railway grant lands in the East Kootenays.

Two future chief foresters, C.D. Orchard and E.C. Manning, accompanied by two district foresters, F.A. Macdonald and E.B. Prowd, in 1932. NA-04750.

He joined the Dominion Forestry Service in Calgary in 1915, continuing to work in British Columbia's railway belt, and joined the BC Forest Branch in 1918, working in Victoria, Prince Rupert and returning to Victoria in 1923. He was appointed assistant chief forester in 1927. But as early as 1922, Manning spoke out about the depletion of timber resources. In his role as chief forester, he took steps to direct forest management in BC towards conservation and the renewal of forest resources, primarily through prompt burning of slash and protection and enhancement of new growth.[11] According to forestry historian Richard Rajala, "never again would the provincial forest bureaucracy be led by a forester so willing to challenge corporate practices."[12]

Manning's support of government intervention led to a sympathetic warning from retired Chief Forester Martin Grainger that Manning should avoid creating "distrust" among those who were suspicious of government regulation. Grainger stated: "Among some people there is a sort of uneasiness associated with you (I can mention that now that you are safely appointed) because they say you are too much tinged with 'New Deal' philosophy, which suggests to this sort of people social insecurity and pink and reddish feelings."[13] Information that Manning provided to the media was used by the Co-operative Commonwealth Federation political party to critique the Liberal government. Although the premier was at times irritated by Manning, Wells Gray (the Minister of Lands) wanted to interfere as little as possible with Forest Branch operations.

The year 1936 saw improvements in the general state of the Forest Branch and forest industry in British Columbia, and Manning began to make his mark. Overall, despite the impact of the west coast longshoremen strikes in the US, the BC forest industry began to see a somewhat hopeful future. A tariff agreement with the United States started opening markets, despite the bumps of labour disruptions, and trade throughout the British Empire was improving.

Manning was able to catch his breath and consider what future forest policy might be once the recovery was underway. He expressed concern over the lack of experienced staff and the gap in bringing along and training young rangers and foresters. In 1935 the branch was still almost 60 staff below its level at the start of the Depression. While the largest toll in terms of layoffs had been on young rangers entering their careers, Manning was also losing his senior staff, due to retirements and competitive salaries elsewhere. Manning emphasized that the Forest Branch was continuing to expand the volume of cut with fewer staff than it had five years previously.

The upturn in the market, increased revenue through the scaling fund and economies in administration had also cut costs. And so by 1936 Manning could see that they were coming close to returning to the profitability that was last seen in 1927.

A new ranger station at Grizzly Lake under construction, 1931. NA-05877.

The Davis raft chained logs into a cigar shape to reduce loss while towing through sinkage or high wind. Selwyn Inlet, 1930. NA-05809.

For Manning, forest conservation included ensuring prompt burning of slash, removing standing dead trees and promoting replanting. He was challenged in implementing these measures during the Depression when the government and forest companies were unwilling to spend funds on forest conservation. In 1936 he went on a trip to inspect logging practices on the E&N Railway belt on Vancouver Island. In this strip of private land – one-fifth of Vancouver Island, granted to the E&N Railway to underwrite its construction – some of the largest companies in the province were carrying out intensive logging with little oversight. Manning commented on the E&N logging practices that "if we continue to do as we are doing now, we don't need to talk about the crop of the future."[14]

Manning recommended to the Minister of Lands that the government increase funding towards research and try to cooperate with industry to protect natural regeneration. He also proposed to the legislature's Forestry Committee that the Forest Branch should have the right to regulate slash burning and logging on lands granted by the Crown. Wary of government interference, companies operating on the E&N lands offered to voluntarily burn their slash if amendments to the *Forest Act* could be postponed. Dismayed by the lack of some companies' compliance that year, Manning pressed for the amendment, asking: "Does the title to E&N land convey to the owner the right to log in a way to leave it barren and a burden to the next generation?"[15] He drew upon the

Victoria Lumber and Manufacturing logging camp no. 5, near Ladysmith, 1907. C-06454.

report that provided the argument for sustained-yield management in British Columbia, F.D. Mulholland's *The Forest Resources of British Columbia* (1937), which revealed a shocking lack of restocking or new growth on cut-over lands. The Liberal government was convinced to pass an amendment to the *Forest Act* in December 1937 compelling companies to burn slash and leave seed trees.[16] The next summer, British Columbia experienced a number of forest fires. The worst one, the Bloedel fire, burned through slash between Campbell River and Courtenay, covering an area over 30,000 hectares, consequently resulting in a budget increase for the Forest Branch.[17]

Informing the public about forest management was a priority for Manning. He worked on improving his public speaking skills and gave talks to diverse groups such as fish and game clubs, women's councils, Boy Scouts and tourism organizations. According to his daughter Helen Manning Akrigg,

> The theme of his illustrated lectures was always the same – that B.C.'s most important industry is forestry, that overcutting and destructive fires leave too much land barren and thus threaten the province's future prosperity. And, since over 90% of our forests are Crown forests (i.e. owned by the people of British

> Columbia), it should be the concern of every citizen to see that they are properly managed.[18]

Forest development projects continued to depend on young, unemployed men. Relief came from the Young Men's Forestry Training Plan, focusing on men between the ages of 18 and 25, and the Forest Development Project for older men, which involved over 21 camps on the lower mainland and Vancouver Island comprising approximately 1400 men. By the second year of the YMFTP program, the previous year's ranger assistants could then be used to help train the 50 men employed at forest experiment stations at Green Timbers, Cowichan Lake and Aleza Lake. Twenty other forest development and protection crews of 10 to 16 men were employed throughout the province.

The YMFTP also enabled a depot-like system for rapid-response fire suppression teams. A graduate forester was assigned to fire control research, and the three organizational divisions of prevention, detection and suppression began to take shape. Experiments with the use of radios and fire lookouts were proving successful and offered an economic advantage, doing away with miles of stringing and hauling telephone wires and improving the speed of response.

Using Baling machine at Green Timbers, preparing seedlings for shipment. NA-06984.

Elk Falls Park in 1947. NA-11760.

Response speed was becoming increasingly important because in the Kamloops district in 1936, for example, a new fire problem was emerging in the large, old pine forests that had been killed by bark beetles. The dead trees, soon tinder-dry and in many areas blown down, formed a bed of flammable material within the summer grazing range for cattle. To remove that hazard and gain more grasslands access for cattle, fires became more common, and some had human help. This relationship between grazing, forest management and pine beetle infestation – overlaid by issues of private property, the commercial value of nearby hayfields, fencing, barns and other improvements – became a complex issue. The impact of the beetle kill rippled through the landscape and affected overlapping economies, creating a significant forest management issue involving multiple stakeholders.

The least traditional aspect of the YMFTP program proved very popular with the public: the development of parks. The Forest Branch and the YMFTP became very active in park development, and the new program began with the Elk Falls Park recreational development work. In 1934 Elk Falls probably received half a dozen visitors a day. But by the summer of 1938 the park was visited by over 2000 cars. With the Forest Branch, formerly unemployed young men improved many parks, small and large. These include Dean Park, Thetis Lake, Mount Douglas, Englishman River, Little Qualicum, Elk Falls, Capilano Canyon, Stamp Falls, Grouse Mountain, Medicine Bowls and Cultus Lake. Almost half of YMFTP work hours were spent on park development. Under the 1939 amendment to the Forest Act, the Forest Branch became responsible for 41 parks scattered throughout the province, comprising more than six million acres (almost 2.5 million hectares).

Exploring New Approaches to Forestry

Manning generally encouraged debate among his staff about new ways of working and thinking about forestry. No one pushed harder than Frederick Davison Mulholland who favoured Sweden's approach. The Mulholland Report, a masterful synthesis of the forest atlas in 1937, argued that what was needed was not just a "crop inventory" and cutting schedule, but a thoughtful comprehensive survey of each forest area, including economics, species, policy and management, silviculture, forest health, cutting policies and protection all aimed at a sustained-yield administration of forests.[19] The discussions were mediated by C.D. Orchard. These new approaches that Manning helped nurture

Douglas-fir near Duncan along the road to Cowichan Lake, 1931. NA-04356.

would emerge via the first Sloan Royal Commission (1943–45) in the postwar Forest Branch. In 1939 the experimental *Coast Islands Working Circles* project was the first use of working circles for forest management decisions, a concept applied more widely in the 1950s. It involved four small provincial forests in the Vancouver forest district: Sonora, Hardwicke and East and West Thurlow forests to provide demonstration areas for testing sustained-yield management.

Manning also introduced a new type of project: cutting plans. The aim was to work cooperatively with commercial logging operations to institute a plan to create the greatest degree of natural regeneration possible within the logging industry. Starting with the 1936 annual report, he promoted the idea of multiple use and stakeholders, noting that

> it is becoming increasingly apparent that we must value our forests not only as sources for supplies of timber, but also for their many other uses – as food and shelter for our game and furbearing animals, as regulators of the water-flow of the streams in which we fish, and as attractions for the tourist and other recreationists who delight in the great outdoors. Forest management plans must recognize both the industrial and recreational uses of forest areas. By farsighted planning we must make these various interests harmonize as much as possible, secure greater returns from our forest areas, and make them the greatest single drawing card for our rapidly increasing tourist trade.[20]

Markets were another issue of concern for the Chief Forester. In 1937 the war between China and Japan virtually destroyed Pacific spot markets for British Columbia wood. Freight prices became unstable. Even shipments to South Africa declined. Despite that, forest products still accounted for 36 per cent of the province's annual income, more than agriculture, mining and fisheries combined. But Manning realized that a new approach to marketing BC wood would also be necessary. He was concerned about the reliance of the commercial industry on Douglas-fir, and he noted in particular the need to expand markets for Western Redcedar and hemlock. Given the proportion particularly of hemlock in the forest, he found the cut rates out of balance with the overall ratio of the forest's species. Notably, on the lower coast he found Douglas-fir accounted for 59 per cent of the cut, Western Redcedar for 13 per cent and hemlock for only nine per cent. Manning was concerned about the long-term implications of that trend. The postwar solution would be the development of markets for hemlock.

Forest Service launches at Nodales, Sonora Island, 1930. NA-04575.

Following the lead of the US Forest Service's successful use of radio in their National Forests, 10 portable radio sets were bought by the BC Forest Branch and installed by G.A. Playfair who spent much of 1937 installing them in selected interior lookouts. With these radios, lookouts in the southern Okanagan networked with their counterparts in the US Forest Service, although some staff had already built their own sets. (One staff member, Bill Lowry stationed in Prince George, built his own a radio telephone to link his office and home). On the coast, radio had arrived earlier and in the 1920s was used to keep contact with the Forest Branch's marine fleet. At that time, the cost of a radio set was equal to the cost of one mile of standard telephone line. The experiment was deemed quite successful, although the quality of communication was not as clear as a standard telephone line. The following year, over 40 radios were deployed by the Forest Branch and saw their first use in fire response camps, keeping the headquarters unit in touch with the forward operations and assisting in the movement of men and supplies.

Innovation in public education about fire included moving picture trailers for theaters, broadcast announcements on radio stations, and newspaper and magazine ads throughout the province. Firms were encouraged to have fire protection messages on their letterhead and on accounts. Merchandise labels also bore fire prevention slogans, and large department stores created prevent fire window displays. Through photofinishing studios, 100,000 "stuffers" pamphlets related to fire education were distributed to students having their pictures taken.

Ranger meeting at Kamloops, 1936 or 1937. F-08413.

Almost 19,000 children and adults went to over 190 meetings in various parts of the province to hear talks, accompanied by motion pictures, about their role in controlling fire.

Education for staff was also very much a part of the Forest Branch's thinking under Manning. A steady stream of skills-upgrade correspondence courses were available during the winter months and sent out regularly. These courses fed, in some cases, into the ranger or scaler exams, or were simply aimed at improving staff's knowledge of law or basic accounting.

In terms of branch communication, May 27, 1935, saw issue No. 1 of the reborn *Forest Branch News Letter* sent out. In those two pages the editor (likely Dick Orchard) celebrated the fact that, claiming to be exhausted, he had managed to assemble two pages, and he promised that this would not be a belated weekly like the 1919–1920 attempt:

> We are interested in Forest Branch work and believe there is enough of real interest and value going on which does not get around the Service, to warrant a letter of this description.

> Every Ranger in the Prince Rupert District has a hose-drying mast. Not just a mast but a damned good mast. Cooper has a beauty. Wilson's is a work of art. Martin put his a mile down the road so it would be easy to lose a few lengths, but then it is a good mast and he lives in an honest community. That is news. The Vancouver Forests District used to have a hose-drying mast. They hung a pulley on a high branch of a four-foot fir. But then they moved the tool cache, for which the tree was used, and they couldn't move the tree. The tree didn't belong to us anyway so perhaps it's just as well they couldn't move it. When some other district develops enough pep to get masts up for every Ranger we would like to tell about it.
>
> And another thing—we are not going to preach. This is positively the first and last sermon which will appear in these letters. Some of the things we have to say may have a moral. If so, you will have to extract it yourself.
>
> Number 1 goes on its way, hesitant and somewhat doubtful of its reception. If there isn't too much adverse criticism, subsequent numbers will turn up "from time to time."[21]

The newsletter was very well received, providing news, humour, advice and shared knowledge to Forest Branch staff around the province. It did much to build spirit and to close the gap between Victoria and the distant ranger stations. Local initiatives were recognized, whether it was a home-made radio system or tips on safer ways to store axes learned from a near miss in a car crash. One suggestion published in the newsletter was that engine oil was best measured in an empty beer bottle. If Victoria wished, the writer suggested, it would be best to ship out full bottles of beer, in order to prevent breakage, and then willing staff would empty them and refill them in the field with oil.

Poetry was popular, as were stories that poked fun at greenhorn errors, such as the man who treed a bear and had a friend keep jabbing it with a pole while he chopped the tree below – until he received the contents of the alarmed bear's rear defense system and spent the afternoon washing in the stream.

The newsletter also reported moves, promotions and retirements. It began to print staff profiles written in a lighthearted manner by the individual. Subjects were selected by choosing fire tags, the metal disks kept at headquarters and issued to each staff person, much like a dog tag. The publication also reported tragic news: the drowning of a Nanaimo ranger's son; the sudden death of a

Mrs Melrose; the death of George MacAfee in Records Branch, an amputee from the Great War; the death of A. Skiba of Cranbrook and W.H. Podavilnekoff of Brilliant, who died fire-fighting in two separate fires. Skiba was one of four trapped at Munrow Lake (the other three had severe burns), and Podavilnekoff was killed by a falling tree in a fire at Salmo, both in 1936. In a much lighter vein, the newsletter also reported the addition of children and marriages in branch staff families.

The lives of the women in the early Forest Branch are almost, but not quite, invisible in the masculine culture. They appeared unidentified in photographs, briefly named as stenographers or as marrying a forester, ranger or a member of the research staff. No staff profiles of women were prepared before World War II. One woman forester left to become a woman of leisure in California, but most were "struck by cupid" or another formed "a partnership agreement," and some simply married. The typical parting gift from office staff was a silver tea service and tray, occasionally an electric cake mixer, and on retirement a travel case.

Miss Irene Aird, the chief's secretary, joined the branch in 1925 and left on medical leave in November, dying on March 10 the following year. Her contribution was summarized as: "Capable, and invariably cheerful and helpful, she set an example by which many of us might do well to profit."[22] In 1936, when the editor of the service's newsletter reported that "a Miss" had applied to take the licensed scalers' exam, the editor stated that the results of the review of the *Sex Disqualification Removal Act* of 1931 were clear: "a she is as good as a he when it comes to a Civil Service appointment. We have had more than one girl on lookouts. Now the scaling staff. Girl Rangers next."[23] That same year, the stenographers of Kamloops wrote in to ask that perhaps the covers of their notebooks could be updated to reflect a slightly more modern view of women than a graphic of a 1920s flapper.

The Great Slash Debate and the Bloedel Fire

One of the great conflicts between Chief Forester Manning, industry and sections of the public was their divergent views over slash, the biomass of debris left behind after logging operations. Operators both small and large saw dealing with slash as time taken from log production, something best put off until later in the season or the next year. The politically active fish and game clubs sided with industry, arguing that slash left on the site reduced soil erosion and provided habitat for game such as deer. But to the forester's eye, slash was simply fuel on the ground waiting for a spark. Compliance was yet another issue, as poorly thought out piling and burning of slash carried its own risk of becoming a source for runaway fires. These issues came dramatically to the fore in 1938.

Smoke bellowing up from the Bloedel Fire, 1938. NA-06425.

While 1937 had been a relatively quiet fire season, the same would not be said of 1938. The number of fires more than doubled to 2412, economic damage rose from $155,000 to $2.2 million, and the cost of fighting fires went from $28,000 in 1937 to half a million dollars in 1938. The reason was evident to all: weather. Almost no rain fell on the Vancouver forest district after May. High, dry winds and low fuel moisture prevailed for much of the fire season that began May 18, with 45 fires the following week.

On July 5 sparks from the engine of a yarding machine in the Bloedel, Stewart & Welch operations started a small fire in a cold-deck of logs by Boot Lake. It was called in about 4 pm, and Ranger A.C.C. Langstroth was dispatched with pumps and lines. He worked with Bloedel, Stewart & Welch's 180-man crew, locomotives with water car and pumps, and a Cat (bulldozer) for cutting firebreaks to plot out how to contain it. Rereading his supervisor D.B. Taylor's staff fire diary, it was just another minor fire, one of several and easily contained. Taylor was aware of Langstroth's work and continued with the task of monitoring and advising on slash disposal at other company camps and sites. Ranger Langstroth watched to make sure the small fire at the Bloedel site was well controlled. On July 7, Taylor checked on the Bloedel site progress by taking a railway speeder over to see how Langstroth was handling it. He considered it to be in "fair shape."[24]

On July 13 Bloedel's Boot Lake fire began to "smoke up." By the 15th, Victoria knew they had a problem. The fire quickly jumped half a mile of green timber. A hot, dry, westerly wind picked up. Four dozen men were sent in with cats to cut fire control trails between green timber and swamps.

On the 18th, at 2:30 in the morning, Taylor recorded a temperature of 78 degrees (26 Celsius), with 46 per cent humidity and hot, dry, variable winds. The fire began to jump roads and trails. Before the next day was out, Taylor was using his Cats to drag other Cats out of the fire. At 7 pm the fire crossed all the trails they had spent the day cutting. He began to lose pumps and hose lines. In the words of the chief forester, the tinder-dry woods exploded "like dynamite."[25] On the 20th the fire jumped the Quinsam River; aerial reconnaissance showed it had advanced 10 miles (16 km) in a day and was moving quickly. The fire rapidly overcame all attempts to control it.

Settlers in the surrounding countryside fled for their lives. The fire was on a scale not seen by BC communities since the Port Neville fire of 1925, which burned 300 square km of forest, 15 per cent in the Sayward Provincial Forest

D.B. Taylor, 1937. NA-06182.

and the rest on private E&N lands. The fire reached the coast and headed down towards the town of Courtenay. The smoke from the massive fire drifted across Georgia Strait, and hung in the air in the streets of downtown Vancouver.

The Bloedel fire drew to it every crew the Forest Branch could find, and all the bulldozers available. Bloedel put 685 men on the fire, and Elk River added their camps after the families were evacuated. The unemployed on the streets of Vancouver were rounded up. The first batch of 200 Vancouver unemployed were quickly shipped off to fight the fire for 25 cents an hour. Others volunteered. The scale of the emergency was such that the Royal Canadian Navy sent three vessels to move supplies and firefighting crews. Sailors were put ashore to evacuate families fleeing in the path of the fire. The navy won high praise for the radio crews they sent ashore to set up emergency communications stations.

July 21 was a bad day. Taylor briefed the mayors of Courtenay and Cumberland about the real danger to their towns. He turned down a plan to back-fire slash as too dangerous. Instead of removing fuel, it might, as he saw it on the ground, just accelerate the speed and size of the fire. Elsewhere that day, District Forester C.J. Haddon led a stubborn battle to save the small hotel at Forbes Landing. His efforts failed, and the hotel was lost. A rumour of the death

A tired crew takes a break from the Bloedel Fire, 1938. NA-06458.

in the flames of Mr Forbes and 150 men who had been firefighting there for a week brought a distraught and weeping Mrs Forbes into the district office. The rumour proved to be false. The Vancouver Province reporter "Torchy Anderson" wrote about growing tensions between the local loggers and local residents, and the tired crews of ill-prepared Vancouver unemployed, who were accused of shirking, cutting six hoses and sabotaging the pump at Forbes Landing by putting sugar in the gas tank.[26] At noon Haddon lectured protesting Vancouver crews, fired 75 and threatened to ship back another 100, who he said were not pulling their weight in the blistering hot work. That view was reinforced on the Friday, when Premier Patullo included it as part of his Friday emergency radio broadcast to the province.

Not everyone accepted the branch's accusations of malingering and sabotage. At least one Vancouver "malingerer" was hospitalized, and the local doctor found the cause: starvation. Some of the crews included street people who probably would not have passed military medical exams. Others pointed out that local loggers had also been idle on the lines, but few were singled out. Another observer commented that when you run over a kinked, pressurized hose with a heavy truck, the line will shear as if cut. At least one crewman did admit to accidentally cutting a hose when attacking a fire with an axe. These instant crews were not prepared for the long situation they found themselves in, without training.

On the 22nd Taylor and a crew scraped the ground to mineral soil for 200 feet (61 metres) and cut trees around for 500 feet (152 metres) to save the Tsolum bridge. It was, however, destroyed by the heat and flames. In an unprecedented move, Wells Gray, the Minister of Lands, used section 119 of the *Forest Act* to order all logging shut down on Vancouver Island except at the northern tip. The fire continued its move south and burned 40 million feet of cut logs; the sound of them going up was heard two miles (3 km) away.

Newspapers added to the desperate tone with numerous stories of the deaths of motherless fawns; bears with burned paws; deer dying in swamps, trapped by smoke and flame; and grouse hovering until exhausted and then diving directly into the flames. Unfortunately, that was also the assessment of conservation officers, whose reports brought the same observations but with a more objective eye. Taylor's own fire diary records seeing several herds of deer, including one of 75, trapped inside the fire. This media coverage deeply eroded public support for fish and wildlife clubs who opposed slash burning.

On July 23 the fire stopped 15 feet (5 metres) short of a dry log storage area containing 30 million feet of logs at the Elk River Timber Company operations, in what *Province* reporter Torchy Anderson declared to be "a Miracle."[27] It was actually the result of the company's locomotive engineer Andy Teak and brakeman "Waggo" Wayne, who had taken their Shay through the fire 15 times to water down the logs. The paint on their engine burned down to bare metal, and they admitted "Well, she was fair to middlin' hot." Torchy Anderson's interview ran under the headline "Sightseeing Trip to Hell."

For the next week, those fighting the fire were lucky. While temperatures stayed high and the fire continued, the wind died right down. But a flare-up in August threatened the community of Bowser. Taylor arrived to find a Mr Hoy had started a back-fire without enough men and had badly organized the few he had. Taylor called in 46 fallers from Courtenay and sorted things out – or so he thought, until he learned the fire escaped briefly the next day because the night crew left the line without replacements. Hoy had decided to fire the day crew without telling anyone. Taylor's fire diary notes that he searched the roads and trails for Hoy all day, finally locating him in the Bowser beer parlour. There are no notes on the conversation that then took place.

Hoy rehired the crew and returned to the line, leaving Taylor to deal with farmers' complaints of damage caused by the Cats cutting fire lines to protect Bowser. Bob Grant, of Grant Logging at Fanny Bay, met with Taylor and told

Crawler tractor working at the Bloedel Fire, 1938. NA-07023.

Part of the area burned in the Bloedel Fire, 1938. NA-06500.

him Hoy had called out his men at 5:30 am, laid them off at 7 am, and called them back at 10:40. Taylor served another formal Letter of Instruction on Hoy, ordering him to hire and keep crews on the fire. Meanwhile, he supervised numerous Cat operators who were proving their worth, despite breakdowns and broken blades. He dealt with an accusation of fraud against the timekeeper at Dove Creek Camp, asked for and delivered a detective, spotted another leftover fire threatening a Japanese fire crew and dealt with that, listened to three "tin-canners" complain about how they were being mistreated, fired them, and moved on to the next problem. And on it went for the 44 days Taylor spent on the Bloedel fire before going to Victoria to prepare his paperwork.[28]

Fallout from the Bloedel Fire

The scale of the fires and the destruction they caused ended opposition to Manning's views on slash and the need for its control. Public and political

Setting up a radio station at the Bloedel Fire, 1938. NA-06431.

support gathered around the views of the outspoken chief forester. The fire diaries and reports were gone over, and the lessons learned were shared with colleagues as far away as California. The Forest Branch learned from the problems with drafted crews and their lack of preparation in scaling up to a large fire's logistics. But they also found positives in the examples of other government employees, like conservation officer Freeman King who took on the specialized task of running a large, complex fire camp. The size, speed and endurance of the fire proved the worth of Cats and the need to stockpile things like dozer blades. Assistance from the navy taught the value of portable, high-quality radio communications and about how to conduct large civilian evacuations. The emergency-purchased new Pacific Marine pumps outperformed the older standard pumps, ran three times longer, used less fuel and had better fittings. Aircraft proved vital to getting a fast assessment of the progress of a fast moving fire.

An often overlooked part of fire is the litigation that follows. Elk River sued Bloedel over the loss of timber and the cost of fighting the fire. The case went to BC Supreme Court and was decided for Elk River. Bloedel, Stewart & Welch's claim for $68,000 in wages from the Forest Protection Fund was denied, since it was their fire. Their legal counsel responded by stating there were two fires – the first they put out; the second being a Forest Branch–lit back-fire that they claimed caused the real damage. Had the ranger given Bloedel legal permission to restart operations after the first? What were the details of the back-fire? All told, Ranger Langstroth spent more time in court and making statements than he spent fighting the fire. Along the way, every weather report, hazard stick and humidity and temperature graph for the region since 1932 was sought. All the lookout reports and staff diaries were scrutinized. Interviews and cross-examinations were conducted, as well as negotiations on what questions could be asked or answered, and everything was transcribed in preparation for the court case. The matter was not finally finished (Bloedel lost) until November 1942, four years later.

Another legacy of this fire was the question of how to deal with its destructive impact on regeneration. Throughout the 1930s natural regeneration was the favoured approach to reforestation. The idea was to encourage companies to leave suitable seed trees standing in cut blocks and to minimize the blow-down of trees surrounding the cutting area. The Bloedel fire not only destroyed any possible seed trees, but it was also hot enough to destroy any existing seed on the ground. Natural regeneration was not an option for the 30,000 hectares. With the backing of the premier, the Forest Branch undertook the largest reforestation project in the history of British Columbia to that point – the first to deal with large-scale reforestation of a wildfire area.

Planting crew at work, 1932. NA-04973.

Early local experiments in reforestation occurred in the 1930s in partnership with Tie Cobb's Elk River Timber Company, but they were small experiments in plantings and spacing work. It became clear that for this scale of replanting, seedlings needed to be produced locally. Ground was levelled and irrigation installed for the Quinsam nursery the year after the Bloedel fire. Relying at first on stock brought up from the southern nurseries, seven 12-man planting crews, working out of tent camps, set to work planting three quarters of a million seedlings in 1939. During World War II conscientious objectors and alternate service workers, including nearby Mennonites, worked out of more semi-permanent camps, continuing the work. By 1951, seven and a half million seedlings had been planted, and reforestation of the devastated area was fairly complete by 1954. The forest that emerged was a point of pride for the Forest Branch and the later Forest Service, a living legacy to those who worked the Bloedel fire and those who reforested the land and managed the new forest.

The Bloedel fire led to some fairly important discussion for the Forest Branch staff about change and policy. The editor of the November 1938 newsletter opened with a thoughtful discussion piece about the future and the lessons of the Bloedel fire.

Bundling seedlings at Green Timbers Nursery, 1940. NA-06981.

Nursery beds at Quinsam Nursery for a new forest, 1941. NA-07654.

> "Day before yesterday" a few hundred men on a fire and a bill for $10,000 cost were something of real concern and a subsequent court of inquiry. "Today" we put a couple of thousand men on a fire and a dozen or more eighty-horse tractors, a couple of aeroplanes and a truck for every erstwhile packhorse, and the cost was sufficient to have turned every District and Victoria head gray under "yesterday's" regime.[29]

The author went on to talk about how multiple use was changing decisions on applications and the need to ask if the cut was in the public interest. The old idea of a few grates for a wilderness campsite now needed mature thought on how to run a large system of parks, an area where the author suspected they were lagging behind the public's wishes. And the author warned against the depletion of old-growth Douglas-fir and cautioned what that meant for hemlock. The service emerged from the 1938 Bloedel fire with its eye fixed on the future.

The Second World War

In 1939 war in Europe once again found the United Kingdom cut off from European supplies of lumber, and Canada became the key supplier of that much-needed resource. Loggers who rushed to sign up to fight were turned back as much-needed war workers in their own right. Wood used for hundreds of aircraft hangars meant faster repairs and saved steel for use in ship construction and vehicle construction. Military camps were wooden cities, and wood was used to repair bombed cities and create gliders for invasions. Not only was lumber needed overseas in Britain, but large amounts were needed to construct military and air bases all across Canada to train Allied pilots and the soldiers Canada would send overseas. British Columbia's forest industry rallied to provide that valuable resource.

At first it was shipping, not wood or labour, that proved the biggest problem. Railcars and freighters were in short supply, and wood began to pile up at sawmills. The priorities for shipping food, men and ammunition across the Atlantic by convoy also created bottlenecks at the Atlantic and at the Panama Canal. Freighters were lost to submarines, and neutral freighters could not deliver to war ports. Solving these logistics required intense intervention by the state.

On June 15, 1940, former Chief Forester MacMillan was appointed timber controller for the Dominion of Canada, with the power to regulate the industry and to coordinate the movement of timber to facilitate the war effort. Two days

The large lumber yard at the Victoria Lumber and Manufacturing Company's sawmill at Chemainus, about 1940. E-00760.

later, on June 17, Chief Forester Manning accepted the post as assistant to the controller for the Pacific coast, headquartered in Vancouver. The export of fir logs was now forbidden, except under a Dominion permit.

As in World War I, airplane-grade Sitka Spruce became a high-priority material for the war. By the summer of 1940 surveys had identified all available stands of high-quality spruce, and camps and mills were again engaged at full capacity. Wooden aircraft, such as the Mosquito bomber and the D-Day gliders, used spruce cut from stands on Vancouver Island and the Queen Charlotte Islands. The needs of war descended upon the forests.

The entry of the United States into the war and the war with Japan put another strain on the system. Of BC's wood production, 40 per cent went to the UK; 40 per cent was used in Canada, for aircraft and hangar construction, factories and shipyards, military camps and even new granaries for a bumper wheat crop on the prairies; and another 20 per cent went to the US for its war effort.

War had suddenly and dramatically ended the unemployment problem. The Forest Branch found itself with a network of understaffed facilities, empty camps and managers without workers. Two-thirds of permanent staff eventually enlisted, both the young and the experienced senior technical staff, and by 1942 demands

G.P. Melrose, 1945. A long serving member of the BC Forest Service, he was eventually appointed deputy minister of Lands. NA-08249.

on staff in the Forest Branch led to the abandonment of several programs. For example, ground surveys, general inventories and the aerial photography survey were suspended. The aerial photographic branch staff members such as Gerry Andrews, with his expertise in stereo-photographic interpretation, were in high demand for assessing bomb damage and invasion beach sites.

Research slowed drastically in the face of war priorities, but it was informal 1941 research that initiated the assessment of suitable Douglas-fir soil sites through the classification of ground, using variables such as moisture, soil and the presence of specific combinations of associated plants, such as Sword Fern habitat. Of minor importance at the time, it was a quiet first step towards an ecosystem-based system of forest type classification.

Materials shortages during the war had other impacts as well. The Thurston Bay repair facility on Sonora Island was replaced with the Forest Service Maintenance Depot, a boat repair facility on the Fraser River. Its machine shop proved indispensable as vehicle, boat engine and water pump parts became unobtainable. The first experiments with light trucks equipped with pumps and water tanks happened in the war years using a diversity of vehicles at a time of parts scarcity. Rangers and administrators became convinced that standardization, something the military taught and insisted on, would be useful. An interesting letter to the Forest Branch newsletter editor from an anonymous ranger thanked whoever had sent them three luxurious new trucks,

From 1914 to 1941, Thurston Bay, Sonora Island, was the main boat facility for the Forest Service's large fleet. Today it is a marine park. NA-04577.

with shocks front and back! The first was "muscovite red," the second "cerulean blue." He noted that the interior of the truck beds differed from model to model. One had a "loose, sprawling over-sized box, which gives the impression that it might start to flap its wings and to fall to pieces prematurely" and used up precious space for interior wheel wells. One of his staff, for reasons not shared, was grateful for the invention of an interior hood latch inside the lockable cab, making it impossible to steal sparkplugs or tamper with the engine. He asked, could they standardize colour and size of the truck bed?

Forest Branch staff trained military servicemen in firefighting and the construction of firebreaks around military facilities. As well, Forest Branch lectures were studied by an officer, then translated and delivered in French to 600 soldiers stationed in Nanaimo.[30] In the far north military were deployed to assist with fires when needed, and Royal Canadian Air Force pilots contributed to fire spotting. Forest Branch boats were made available to the Royal Canadian Navy for coastal patrols, and an alliance was formed with the Vancouver Island Power Boat Squadron, providing civilian support for transportation of firefighters trained in emergency evacuation work.

The Forest Branch's weather station work became critical to supplying information to the air force, navy and armed forces. But staff had to learn to

Forest Service launch *Cherry II* on ways at Marine Division, 1946. NA-08764.

laboriously convert the weather reports into code before radioing them in, in case of interception by Japanese submarines or aircraft. Lookout staff along the coast also trained in aircraft identification.

To meet initial needs in the first year of the war, portable radios and base stations were loaned to the armed services. But by 1943 the service had been resupplied and had over 200 radios in operation, as well as almost 300 vehicles, 20 bulldozers and graders, 38 launches and 20 railway cars. Yet all was not well. Batteries and tubes for radios became difficult to obtain. New orders for radios and vehicles remained unfilled. The largest worry was the destruction of a factory during the Coventry air raid, as it was the sole source for specific engines and parts for the motors of a firefighting water pump. Parts for outboard motors also became difficult to obtain. Private repair shops were booked up with war priorities, and staff had to rely on their own skills and informal network of contacts within the Forest Branch as motors were cannibalized and second-hand outboards were purchased to keep things running.

Alternate Service Workers and Conscientious Objectors

Fire, forests and labour shortages led the province to hold talks with Ottawa about what to do about fire suppression for the forests of British Columbia.

Alternative Service Workers camp, 1942. H-04971.

The answer was found in the labour of alternate service workers. Most were conscientious objectors, classed as medically fit for compulsory military service but who refused military service on religious grounds or personal convictions. In such cases, the individual offered to perform non-violent alternative service work. This was the choice of many Mennonites, Quakers and individual pacifists.

With the signing of the Dominion-provincial agreement in 1942, 1000 alternate service workers scattered across the province in forestry work crews. In May, the first of 1000 alternate service workers or conscientious objectors arrived on the coast of British Columbia for work in fire suppression. They would form the backbone protection of BC's forests against fire for much of World War II.

The Green Timbers nursery served as a central reception camp as people were transferred to and from there. After initial processing, they were sent out to one of 20 camps established in high fire risk areas along Vancouver Island and the lower BC coast. Most of their work involved regular fire protection duties.

Outside of fire season work, the crews were involved in falling snags, ditching, laying telephone lines, cutting trails, removing abandoned logging railways and converting the road grade for automobiles, building firebreaks, removing fallen trees from highways and planting seedlings for the reforestation project. Similar programs took place involving road construction, such as building the Big Bend highway. One mobile 10-person team specialized in park upgrades. They built change houses, campsites, staff buildings and trails. They also dug garbage pits, fire pits and latrines; made signs; planted shade trees in parking lots; and cut firewood for campers. In the winter many of the small ASW fire camps closed, and the workers were gathered at a few large camps for larger-scale forest improvement projects.

The alternate service workers impressed those who worked with them. They were confident, calm and motivated. But the alternative service program wound down almost as suddenly as it started, as Dominion funds ceased in March of 1944 with the war's end in sight. As well, British Columbia had been the primary beneficiary, but now other provinces asked that alternate service workers be assigned to work in their home province. The sudden end of the program, just before the summer fire season, also brought reforestation to an unexpected halt, and that in turn meant the nurseries could not ship out seedlings, which were retained for a second year.

For fire suppression work, there was a brief attempt to replace the alternate service workers with 16 crews of high school students, arranged in seven-person fire suppression teams with a foreman, cook and vehicles. This had mixed but generally positive results. There was a problem attracting quality adult supervisors, cooks and foremen willing to take on short seasonal work. Each group had as much advance training as the region's weather allowed and a light vehicle for rapid response.

It was in the Kamloops and Nelson forest districts during 1942–43 that labour shortages and no access to alternative service workers meant they began to perfect the smaller, truck-ready rapid response crews of eight or five people. The result proved effective, providing a fast, ready, moving-in-two-minutes response that dealt effectively with a higher number of spot fires after lightning strikes. Older fire crews had problems adapting to the rapid response, but not the younger team members. And so the focus shifted to recruiting young crews with an experienced leader. This experiment was watched closely as a new model for fire response.

Despite the loss of many technically trained staff, the Forest Branch undertook small but significant research during the war years. There were experiments on working circles, mensuration improvements, continued

Filling a water tank mounted in the back of a truck. NA-07009.

Radio at Nanaimo for a standby fire crew, 1942. NA-07738.

standardization of procedures and increased productivity from seed nurseries, including ongoing disease studies. Studies were done on the Engelmann Spruce (*Picea engelmannii*), a primarily interior subalpine species often found in drier areas with Lodgepole Pine. Experiments with selective cutting, rather than clear-cutting, were conducted in 1941 to see if natural regeneration could be improved. Other studies examined patch cutting. Smaller, lighter machinery was examined, in order to try and reduce damage in yarding and to make it more efficient. A new Forest Branch nursery was set up at Duncan. Crop studies continued, and techniques for embedding Douglas-fir seed in peat or moss were tried (but found ineffective) as the search for an improved growth medium for tree-planting continued.

The war also brought about improvements in radio technology. The battery powered Radio Specialty Company's SPF Portable Radio Type AE had, at 2.5 watts, an official range of 25 miles (40 km), but in practice could reach six times that distance, to the surprise of the maker and the US Forest Service Radio Laboratory. The Forest Branch's mechanical supervisor visited both in May 1939. At the RSC factory, he looked at the new "lunch box" kit, a 6 lb (2.8 kg) unit useful for short-distance communication in fire and field operations.

Radio Culture

Radio allowed staff to send and get messages home, and changed life in other ways. The Forest Branch newsletter gives the odd glimpse of how radio changed the isolation of life in the lookout tower:

A Radio Conversation Not Entirely Imaginary:

This is VJX6 calling VX7Tay….I mean this is V6Jay Tay calling VSEXT7, I mean this is... Oh Hell, I mean this is Vx6T *calling V7TT T, THIS IS CRAG MOUNTAIN Lookout calling Peak Lookout.*

"Hello, that you Bill? Calling Peak Lookout. Come in, switch over will you. Hello, that you Bill? Say do you know....Hey, get out. No, not you Bill. I was going to ask…GET off the air. *What? No. Say, Bill, I was…Get off the air. Some blanketty outfit Bill trying to butt in. Yeah, some other District. Want the air all the time. Look Bill, do you...hey...Foghorn Lookout. Get off the air.*

"There goes somebody else. Hey fellows, let me talk will you. I've got an important message to get through. Hey, gimme a chance. Hello, Bill. Still on. That you Vay jex 7T...that you Bill? Say listen Bill...there goes that blasted Foghorn Mountain again. Say, I've got an important message to get through. Yes, that's right, get off the air...

"Hello Bill, everything seems quiet now. Listen Bill before they get going again. How do you cook *rice?"*[31]

The powerful S-50 transmitter added range to communications. I-32049.

The mechanical supervisor's responsibilities included discussions with the USFS, visits to their warehouse, and looking at new equipment, like the new Edwards firefighting pump, a new power saw, the experimental "kitten" trail-breaking bulldozer and supply carts. He was even tasked with the search for a better vehicle oil filter.

An 800-watt S-50 base station at Victoria covered most of the province, while each ranger station had its own PAC receiver/transmitter base station, with a radius of 150 miles (241 km). The use of UHF (ultra high frequency) radios by the armed forces was viewed with great postwar interest by planners in the Forest Branch as a clear direction for limited-range, static-free, portable ground communications. They became convinced about the use of UHF radio when in 1943 a UHF set was used to restore communication at the Nelson district Saddle Mountain lookout after it took a direct lightning hit.

The increased use of electricity in urban areas around ranger stations was beginning to create problems with radio interference. As an experiment, in Prince George telephone wires were spooled out of town from the courthouse to a remote transmitter that could be turned on and off remotely. The result was a static-free reception and transmission. By the end of the war the radio centre at Victoria headquarters was receiving hundreds of messages and gathering weather reports daily from across the province by radio to calculate fire risk.

The Ranger

At the centre of the culture of the Forest Branch and the Forest Service is the ranger, that singular figure of masculinity, a leader of men who conquers fire, who hires and fires, finds lost hunters and children, supports community needs, and one who can stand toe to toe with angry loggers (and bears). The ranger personified "common sense" government, there to protect forests, range lands and rural communities. No one believed in that ideal more than the Branch staff who wanted to be rangers themselves. The ranger continues, to this day, to inspire Forest Service staff with the importance and responsibilities of the role they play in rural British Columbia.

Ranger D.B. Mellville crossing the Nemaia River, 1933. Mellville was stationed in the Alexis Creek Ranger District. NA-05969.

The Ideal Ranger

"A couple of 'Recipes for a Ranger' put in here on account of we are so fed up with thumbing them over on the file we think we had better print them and get rid of them." – The Editor, *Forest Service Newsletter*, May 29, 1937

"Recipe for a Forest Ranger" by M.K. Schoder

Take a pound of ambition and mix it with tact,
Add initiative, sand, and of courage no lack.
Mix well with frankness, and season with grit,
Add a large pinch of pep, but pepper omit.
Put in energy, honesty, steadfastness of goal,
With a sense of humour, then garnish the whole;
Now place in the kiln of experience hard won,
Don't leave him half-baked, but be sure he's well done,
And out comes your Forest Ranger!

"Recipe for a Ranger"
First get a big kettle and a fire that's hot
And when everything's ready, throw in the pot,
A doctor, a miner, of lawyers a few;
Add one sheep herder and a cowboy or two,
Next add a surveyor, and right after that
A man with good sense and a good diplomat;
At least one stonemason, then give it a stir,
And add to the mess one good carpenter;
A man that knows trees, and don't leave from the list
A telephone man, and a fair botanist.
The next one that's added must be there, that's a cinch,
It's the man that will stay when it comes to a pinch,
Add a man that will work and not stand and roar,
Who can do ten thousand things and then just a few more,
Now boil it up well and skim off the scum,
And the Ranger you'll find is the residuum.[32]

However, in 1939 Ernest Manning foresaw changes being needed to the old ranger system. The future would be a very different world, one of technical complexity, based around concepts like multi-use stakeholders and sustained yield forestry, and a very data-centric world. Would the future still have a place for the wild frontier legend of the early Forest Branch ranger? He discussed it at length in the 1939 *Annual Report*:

> The Ranger staff constitutes the back bone of the Service. On them devolves all the routine field-work of general administration. The service depends on the Ranger for timber estimates and appraisals, contract conditions and upset prices for timber sales, checking of cut, and ensuring compliance with regulations and contract conditions on all cuttings; prevention and checking of trespass; grazing administration; organization and supervision of protection organization and direction of active firefighting; together with a host of minor duties and direct contact with the public. The success with which they

have discharged these varied and responsible duties over a long period of years is worthy of the highest commendation.

We have no record of how these men were selected in the early days of the service, but in 1919, a system of selection by competitive examination was instituted. Early examination records, too, are somewhat sketchy, but tracing back through the sixteen years 1924–39, we find an unbroken record of strict appearance to the principal vote of appointments on the results of these competitive examinations. During this period twenty-five open Ranger examinations have been held at which 287 contestants were examined. One hundred and forty-six men qualified, fifty-one of whom received appointments in order of merit. The result has been an exceptionally well-qualified staff.

The Ranger job, however, has developed into something quite different from what it was twenty years ago. Management concepts and Forest protection methods have made great strides. Technical factors unthought of in past years assume first importance and the so-called practical man in the future will be unable to cope satisfactorily with problems he will be called upon to deal with. Existing Ranger staff will, for the most part be able to carry on by virtue of long experience and field training. For new recruits, we find it necessary to turn more and more to men with a background of technical education. Developing along these lines, five technical foresters have been drafted into the Ranger staff during the past three years. This year ten forestry students were employed in temporary capacities in the part-time forest protection staff, with a view to training for the Ranger grade at some later date, and it is proposed to continue this system of employing student assistance. The probabilities are that in the near future it will be necessary to demand a forestry degree as a primary qualification for the Ranger grade appointment, or to introduce technical papers in the Ranger examination, or both.[33]

F.A. Macdonald, 1933. NA-05900.

Ranger Frank MacDonald
Frank came to BC from New Brunswick in 1898 and worked at Thurston Bay in May 1917. "Mac," as his friends called him, held Ranger Certificate No. 1 and went on to be supervisor; assistant forester (Victoria); district forester (Nelson); assistant forester and forester in operation (Victoria); and assistant district forester (Vancouver). When he died after an illness the editor of the *Forest Branch Newsletter* said of him: "One of the best known and best loved men in the Branch his memory will be green with us for many years to come."

The Death of Manning

As a war measure, in July 1940 a timber controller was appointed for Canada, and Ernest Manning became the controller's assistant on the Pacific coast. He threw himself into the work with the same energy that he approached his administration of the Forest Branch. He died in a tragic air crash in northern Ontario on February 6, 1941, while returning from a business trip to Ottawa. A shocked Forest Branch gathered to bury their second chief forester to die in service. Manning was succeeded in the role of chief forester by C.D. Orchard.

Writer and Naturalist Roderick Haig-Brown on Ernest Manning:
E.C. Manning, in five short years as Chief Forester, had won the public confidence in a way that few civil servants can or do… He gave the citizens of the province facts and figures that they could understand, he told them the state of their affairs, that it was a bad state and that it could be made better and that he would make it better. Because he was frank and fearless, they trusted him and gave him support… He was able to secure the passage of fire prevention measures that should in time make a great difference to forest regeneration. He was able to lay some of the foundations of sane forest policy… He was able to develop the beginnings of a system of Provincial Parks that may one day be really worthy of the province. And he brought the whole forest service into better repute throughout the province than it had ever known before.[34]

Perhaps the greatest legacy the province and Manning's peers could give him was the establishment of the 70,000 hectare E.C. Manning Provincial Park located, 26 km east of Hope, BC, on the road to Princeton, BC.

Further Impacts of the War: Parks, Planes and a New Name

The war had unexpected impacts. A symbol of friendship between Canada and the United States, the Peace Arch Park became part of the Forest Branch's park program. While travelling restrictions created a dramatic drop in foreign and US visitors to Canadian parks, the same travel restrictions led to higher local park use by BC citizens. Alternate service workers worked on a wide range of parks projects, from construction of small park buildings and registration booths to creating picnic tables and ski runs, converting old logging roads into better fire suppression or access roads, and working on overall maintenance and preservation of many of BC's beautiful parks. Only the most distant and isolated parks, such as Stamps Falls, saw dramatic drops in attendance. As a result the war saw two staff appointments of park rangers. For the first time Forest Branch staff were assigned to work full-time on the administration and development of specific parks. One went to Wells Grey Park and the other to Mount Seymour. The park system was organized into Class A, B and C parks. Under the separate Parks Act, 49 parks were the subject of considerable improvement throughout the war and continued public use.

By 1943 over half of the Forest Branch, or 156 members, had enlisted, and 11 more followed before war's end. A.J. Leighton, a clerk from Prince Rupert, and J.H. Benton of surveys division became the first two of nine known staff fatalities in the war. As a result, the fire season of 1943 in the Fort George region's north proved almost impossible to manage. Fires that could have been extinguished grew out of control due to the lack of float planes to bring in crews because of the shortage of aircraft. Headquarters staff began to consider what role aircraft would hold in the future of northern interior firefighting. For the area north of the Grand Trunk Railway, only aircraft were capable of spotting the small spot fires. To quote Shakespeare's *Henry VI*: "A little fire is quickly trodden out, which being suffer'd, rivers cannot quench."[35] Only aircraft could rapidly deliver what the Forest Branch now saw as the future: two- or three-man fire suppression crews needed to extinguish spot fires. In 1945 Fort George district also pioneered the idea of seasonal stand-by charters, with two float planes on call for fire reconnaissance and detection, plus delivery of crews and supplies, as needed. And for the first time the annual report began to discuss the potential of a new form of air transportation: the helicopter.

The oddest story of the Second World War concerned the secret attack on the west coast of North America by Japanese high-altitude fire balloons, filled with incendiaries and explosives. From November 1944 until April 1945, 9300 high-altitude, hydrogen-filled balloons were launched up into the jet stream by a military unit on the island of Honshu, Japan. The goal was to randomly sow panic and set forest fires in North America. Of the 294 balloons known to reach North America, 101 were found in Canada. The first was found in Saskatchewan, and eventually 57 landed in British Columbia. They caused very little damage, as less than 10 per cent that landed actually detonated. Most of the balloons arrived outside the dry fire season, except one at Dease Lake, June 27, 1945. No news was broadcast about the attacks, not even about the death of a family of six in the United States, the only fatalities resulting from the Japanese balloons.

By war's end Chief Forester C.D. Orchard had great reason for optimism. Forestry staff began to return to positions left open for them when they enlisted, and forest revenues were at an all-time high. It was expected that postwar rebuilding would, as before, ensure a steady market. New technologies and new fire suppression teams had been considered and were now ready to be deployed. Replacements for tired vehicles and pumps were assured, ending maintenance and parts headaches. Air and ground surveys, suspended for four years, could now be reintroduced. Cooperation with other agencies was high, and the USFS provided smokejumpers and a plane to assist the Kamloops district in one cross-border fire.

Float plane on Stony Lake, 1948. NA-12658.

The reorganization of the Department of Lands into the Department of Lands and Forests, with a separate Deputy Minister responsible for forests, seemed opportune. On April 5, 1945, the Forest Branch, now with its own Deputy Minister, became the Forest Service. As well, the Royal Commission on Forestry report by Chief Justice Sloan was moving towards it release. Begun in December 1943, the last public hearings ended in July 1945. A clear forestry policy for the postwar world was about to be presented to the public.

Chapter 4
The Orchard Years: Striving for Sustained Yield

Ernest Manning's death in 1941 triggered the appointment of Assistant Chief Forester Chauncey Donald Orchard to the position of chief forester. Manning had popularized the issue of how British Columbia's forests should be managed, but Orchard was the architect behind its shape and implementation. He led the BC Forest Service for the next 18 years, and his ideas on forestry and forest management became firmly entrenched into the organization's procedures and actions. For over a decade he almost singlehandedly determined the Forest Service's role in British Columbia. But he needed the support of the department's ministers to achieve his goal, and each minister did publicly support the sustained-yield policy as it was adapted from the first Sloan Royal

C.D. Orchard standing in front of the BC Legislative Buildings, 1932.
NA-04302.

Commission, given force by subsequent legislation and then firmly reinforced by the second Sloan Royal Commission.

Sustained yield was not a new concept. It was a well established idea in parts of Europe and widely accepted by trained foresters in North America, but it still very much reflected an era that viewed forests as a resource only. Sustained yield was not about preservation; rather, it treated the forest as, in effect, a farm from which to draw a perpetual crop. Trees were to be harvested, turned into logs and used to create wood products. When Orchard drafted his approach to managing BC's forests, this was the accepted goal of most foresters. As N.V. Basnett, one of the leading international foresters of the 1940s and '50s, told the British Commonwealth Forestry Conference in 1952,

> When such a forest (a previously unmanaged forest) is opened for exploitation the primary aims of wise management are to harvest each year from the existing growing stock a volume which will be economic to cut, and meanwhile to procure regeneration and bring to maturity, so as to enable the same, or preferably, a larger volume to be cut annually for ever.

Orchard was determined to make British Columbia a leader in sustained yield. But his vision was not limited to harvesting, even though his first goal was to manage the forests for economic reasons. Rather, he also looked to the

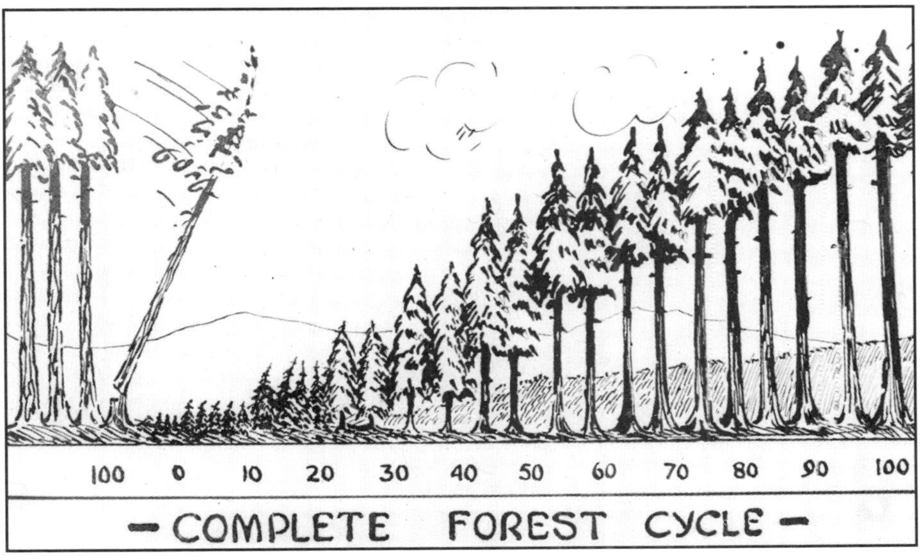

A normal forest under sustained yield showing the various age classes from seedling to harvestable timber. I-52784.

forest as a place for all British Columbians. Orchard's support for parks and the preservation of recreation areas demonstrates that even when preoccupied with creating a stable business organization, he sought to enhance other values, although not at the expense of industrial development.

Convincing the People

Orchard was educated as a forester in New Brunswick before he came west in 1920 to join the BC Forest Branch. He was initially based in Victoria as a forester, engaged in survey and inventory work. He returned to university and in 1926 received his Master's of Science in Forestry from the University of New Brunswick (UNB). Following his return to BC, Orchard was appointed district forester for the Fort George district in 1927, where he remained until 1932, except for a short interval as district forester in Cranbrook. He then moved to Victoria and was appointed forester in charge of the operations division, until his appointment as assistant chief forester and, subsequently, chief forester and deputy minister. Until Dr Orchard (who had been granted an honourary doctorate by UNB in 1947) retired from the BC Forest Service in 1958, he focused on developing a sustainable forest industry. Under his direction, nearly all Forest Service activities would be directed toward that single goal: the creation of a sustainable forest supporting a sustainable industry. The objective was to ensure old-growth timber would be available until a new forest had grown and was ready for harvesting.

Manning and Orchard had convinced the government that in order to save British Columbia's forests, sustained-yield forest management was essential. As others had before him, Orchard publicly announced, in 1942, that British Columbia did not have the extensive timber resources that many thought and that the resource was being harvested at an alarming rate. Immediate measures, he said, must be taken or the industry would fail within decades – and with it, provincial revenue would also drop significantly. This message was taken up by Lands and Forests Minister Edward T. Kenney, who was convincing and forthright in his views. The recommendations of a Royal Commission also provided credibility for new legislation to establish a new policy framework. Orchard fought to make this policy a success and had the strong support of his department's government ministers, especially Gray and Kenney. In 1951, defending the policies now in place, Kenney said to the Advertising and Sale Bureau of the Vancouver Board of Trade:

> How far can we go in over-cutting our great forests in order to supply the insatiable appetite for British Columbia

Lumber throughout the world? When do we start committing economic suicide?

It is almost unbelievable to me, but even though we have been overcutting from 10 to 30 percent for the last decade, the tremendous virility and expanse of our forests have been able to absorb this punishment so that we are still not on the verge of calamity. British Columbia has not slipped to that fatal point in

Lake Logging operations on the Upper Gordon River near Cowichan Lake, 1937. The Forest Service was always concerned about wasteful and destructive logging. Enforceable regulations were needed. NA-06291 and NA-06292.

> forest depletion where the balance of nature has been thrown out of adjustment once and for all. In other words there is still hope.[1]

Such support gave Orchard the ability and authority to implement his vision in British Columbia – a vision that changed very little over the years and that could now be put into action by a revitalized organization.

Immediately on the close of World War II, and with the submission of the Sloan Royal Commission report, new legislation in 1947 moved the Forest Service into action. A clear agenda emerged: training the new staff for new roles, getting the message out about new regulations and then enforcing the new regulations and legislation. The task of getting the message out was critical to success. Industry had to be informed of the regulations, and public support was essential. Getting information to the public was the job of the minister, the newly created public relations and education division, and the local staff in the districts, although Orchard also occasionally spoke on public occasions. Minister Kenney, too, regularly presented the philosophy and the practicalities behind the new legislation. In his radio address on CBC, Vancouver, on March 15, 1948, Kenney stated his complete support of the Forest Service's position of the late 1940s. The forest was a crop and needed to be treated as such, but instead of a one-year or even a 10-year rotation period, it had a 100-year rotation period, and Kenney acknowledged that "we cannot enjoy the maximum forest income without long-term planning."[2]

The enlargement of the public relations and education division was considered to be a vital aspect of the new program. Although the division continued its campaigns regarding wildfire protection, it expanded its work

T.C. Jones showing films to a school class, 1956. NA-16500.

L. Moutray showing films to a service club, 1958. NA-17149.

in other areas of concern. One of its new goals was to convince British Columbians, and especially industry, that sustained yield and the new regulations were essential for the survival of the forest industry. They sought to convince British Columbians that the Forest Service was managing the Crown forest in the best interests of everyone. By using films and photographs, creating brochures and posters, preparing news releases and holding meetings, the Forest Service convinced most British Columbians in the 1950s that the path to economic prosperity and good forest practice was through the government's sustained-yield program.

Going to School

The Forest Service's new and expanding sustained-yield programs required a retrained and expanded staff. Many staff had left for war service and were only now returning to the fold, after years away doing other work. New laws, new regulations and new harvesting methods were all coming into play. One of the big questions was where the trained men were going to come from.

Assistant ranger examinations being held in Langford, 1958. NA-17120.

Certain basic knowledge and ability had always been a requisite for advancement. At least as early as 1921, prospective rangers wrote examinations demonstrating specific knowledge and skills. Industry, with the introduction of sustained-yield policies, also quickly realized that they needed to employ trained foresters, and what better source could there be than the BC Forest Service? Here was a cadre of trained men with an established rapport to those who remained working for the government. And the government, even in this expansive climate, could not find or afford to hire only university-trained foresters. The ranger school was seen as the solution.

Classroom at the ranger school 1947. Left to right: (front row) H.V. Hopkins, H. Barker and H. Stevenson; (second row) K.A. McKenzie, A.F.W. Ginnever, L. Chase and J.F. Killough; (third row) N. Threatful, A.F. Specht, C.L. Gibson; (back row) J.S. Macalister and W. McCabe. NA-08963.

Forest Service ranger school in Surrey, 1952. NA-13701.

The Forest Ranger School, organized in early 1945, became the backbone of Forest Service training for rangers and assistant rangers. In order to institute the program as quickly as possible, the school was initially housed in the old relief and alternate service workers' camp at Green Timbers, Surrey, although new buildings were soon in the works. The school remained at Green Timbers for its whole existence.

The first class opened on January 7, 1946, with a formal opening statement by Minister Kenney, and 20 students began their studies. This first class included two rangers, 10 acting rangers and eight assistant rangers. It was a tough program of six months, divided into two three-month sessions, and in 1949 it was expanded to nine months. The courses covered a tremendous range of topics, from fire suppression to legislation, scaling to office methods. The second class inducted was also heavily weighted with rangers. Of the 20 students in 1947, 12 were rangers or acting rangers. This composition would gradually change over the next few years, and the school moved instead to become the training ground for assistant rangers who forest district management thought had the potential to become rangers. Two permanent staff appointed to run the school provided most of the training, with R.D. Greggor, formerly district forester for the Fort George forest district, as the school's head. (In 1952 Greggor left supervising the school in order to take charge of the newly established engineering services division.) The school drew heavily on specialized staff for lectures, workshops and other instruction and used a variety of outside agencies for their expertise as well, including the federal government for entomology and forest pathology, provincial police officers for instruction on law enforcement and the St John's Ambulance Society for first aid. Industry also helped. For example, in 1949 both British Columbia Forest Products and the Victoria Lumber Company provided assistance for field training. The school also assumed certain specialized training when required. For example, training was instituted for Vancouver forest district lookout men. When not occupied by classes, the school's rooms were often used to hold larger meetings, such as district forester meetings and Vancouver Forest district ranger meetings.

The student selection process was revised in 1951. Most of the men with some experience and already in ranger positions had been through the program, and now the school was training more and more employees near the start of their careers. The teaching staff initiated tours of the province and interviewed prospective candidates, after which they made their recommendations to the various district foresters.

Unit IV, class IV, June 7, 1976, at the ranger school. NA-32401.

The instructors, now with five years' experience, decided to extensively revise the teaching program to make it more relevant. The new 1952 program, for example, included 30 hours of training in public speaking and courtroom procedure. In 1953 two more subjects were added: ranger district organization and public relations. In 1974 the advanced ranger course was replaced by a new training program known as Unit IV, and in 1978 the last of the ranger classes was held.

Renamed the Forest Service Training School in 1962, the school continued to function until 1981, when it was subsumed into the Personnel Branch and soon disappeared. Between 1978 and 1980 the school provided specialized courses before being phased out. It continued to operate into the 1990s, in its later days as private firm called Lorax. Although Lorax continued to operate until the late 1990s, they moved out of Green Timbers in the late 1980s. Increasing numbers of trained students were being produced by the colleges and technical schools, such as the BC Institute of Technology. As well, a major Forest Service reorganization in 1978 resulted in the elimination of the ranger position. It was deemed that a formal school was no longer required, although special training would continue as needed.

Creating the Forest Service

Critical to the success of the Forest Service's new direction was to remake the organization into one capable of implementing the massive changes in how British Columbia's forests were administrated. As a result, the late 1940s and early 1950s were a time of great change. Many long-serving personnel, who had started working in the 1920s and even earlier, were retiring. At the senior levels, this created some options for an easier transition to a new organization. The two most prominent departures were E.B. Prowd, who was responsible for the management and economics division in early 1951, and Assistant Chief Forester R.C. St Clair, who retired in early 1952. The nature of the workload also changed dramatically, as did the way funds were allocated with the new priorities. Small divisions were suddenly to become large divisions, and completely new divisions were organized. Occasionally, activities that had been scattered throughout the service were consolidated, such as when the public relations and education division and the engineering services division were created. In 1952 the Forest Service created two large branches: Technical Planning and Operations, each under an assistant chief forester reporting to the deputy minister. The offices of

C.D. Orchard making presentation to assistant Chief Forester R.C. St Clair on his retirement in 1952. NA-13456.

deputy minister and chief forester remained in the hands of one individual until C.D. Orchard retired in 1958.

Orchard kept authority very centralized; his approval was required for every major and many minor decisions. Victoria headquarters began to control many of the Forest Service activities once under the jurisdiction of the district foresters and rangers. The authority to implement policy remained in the field, but policy and regulatory direction were now set by the office in Victoria – although with regular consultation with field personnel, especially the district foresters. Administration of even the smallest details were now controlled by the Victoria office and the new sustained-yield units, especially the working plans division. As an example of Orchard's involvement with small details, he circulated a general memo indicating his concern regarding the BC Forest Service badge and how it could be modified. On February 4, 1949, Orchard wrote to all district foresters and division heads that

> There will be circumstances under which it may be desirable to add to our Forest Service Badge a distinguishing title to

The BC Forest Service issued a variety of badges over the decades, but a central feature in most was a tree. Later a wide range appeared to designate the various Divisions and Branches of the Forest Service.

> indicate the branch or type of work on which a Forest Officer is employed; for example, Parks or Management. Approval for any such addition to the badge is to be obtained from the Chief Forester before being used: and in the event of approval, the design is to be as illustrated in the accompanying sketch.[3]

Orchard kept very close tabs on all activities until Ray Williston was named minister in 1956 and began to appropriate some of Orchard's authority. Whereas Orchard had previously managed to generally set policy, as well as assume the detailed administration of the forest program, Williston, following his appointment, gradually positioned himself to take on Orchard's role.

Staffing a Typical Ranger Office, in this Case at the Fort St James Ranger District

Bob Perchie: Retired Forest Service Officer

Well, your last question was: who was at the ranger station? There was a ranger in charge, Bill Brash. Then there were a number of assistant rangers. And they were the actual field people that went out and did the various field events. And then there was a dispatcher, who was me. And then in the summer months, there would be at least one lookout person, because above Fort St James there was the Pinkut Lookout. A person was up there for the summer.

And if it got extremely hot, they had secondary lookouts, which may or may not have somebody going up. But there was Ray Westlake, Dave Rowe and Keith Blom, are the three that I remember as permanent employees at Fort St James.

Ranger station at Sechelt, in the standard design of the post-World War II period, 1950.
NA-10421.

At that time – and remember, I was new to the Forest Service – what was known as district office, which is now the regional office in Prince George, was decentralizing their cruising crews and their silviculture crews.

Now, the cruisers went out and measured the timber and measured the volume that was on the ground. After that was done, silviculture crews would come out and they would walk through the bush determining the type of logging that might take place in there. At that time, Prince George was known as the spruce capital of the world.

A common practice was they would leave seed blocks. If you had a 100-acre block of timber that was going to be logged, throughout that block of timber they would leave five-acre blocks of standing timber. The theory was logging would take place all the way around it, and then from the matured timber that was left, the seeds would blow into the logged-over area and regenerate.

Each of the new divisions and branches were critical to implementing the new program. Engineering built the ranger stations and field offices required, plus roads and bridges, and they also surveyed access. The small loggers working in the Public Working Circles (sustained-yield forest management units) in the 1950s needed help if they were to provide the economic prosperity the government of the day sought, and roads were critical to that process – so critical that a separate fund was established to ensure the program had sufficient resources. Surveying and inventory work also expanded dramatically, in order to provide the necessary information about what these new sustainable forests comprised. The forests could not be managed if the Forest Service did not know how much timber was available for harvesting. In order to make the forest sustainable, silviculture had to develop to match industrial expansion.

Growing Up in a Ranger Station

Michael Blackstock is a second-generation employee who grew up on a series of ranger stations in the 1960s. We asked him what it was like.

Basically, my first job was picking cones when I was in grade 2 with my mom. Like, we're going out to squirrel caches and filling up. I remember filling up these kind of hemp bags, and we were getting money. It was cool, you know? It was great, kind of making money that early. So there was always a connection, I guess, between the ranger station and work for me. Picking cones – we did that. Lots of different summers we were out. We'd go out to sites and pick cones. Firefighting, I remember, you know, dad being busy all summer long, and we'd have to.... Our holiday plans would change based on fire season. We didn't know if we were going on holidays or not. And all the excitement around the ranger station. The activity around firefighting was always very exciting as a kid – helicopters and tons of people coming into town. And we were also doing the firefighting at house fires. It was the local volunteer fire department, usually. So there's a very close connection between the Forest Service and communities.

We were the fire department. We were ... helping people out with their education. A lot of volunteers, they would be the volunteers for other organizations, and we'd have organizations like Rotary and those sorts of things. They were the summer employment for the community, either cone picking or firefighting or whatever.

Rescue. They were the people that were called to rescue. I remember in Summit Lake there was a boy that was lost on the lake, and it was the Forest Service that was called to go out. They have the equipment, they have the mobility, they have the expertise in the bush. So they were firefighting, search and rescue, tutoring, leaders in the community. So all of that.

A.H. Bamford: Assistant Forester, Reforestation[4]

Planting projects are normally organized around a portable camp with accommodation for 75 men. The camp is run by a foreman and a planting supervisor directs the work of the three or four fifteen-man crews. Such a unit can plant 1,500 acres during the relatively short six-week Spring season. A good planter can average from 700 to 800 trees per day, depending on the size of seedling and type of ground. The method of planting is fairly simple. A grub-hoe or planting tool is used to dig a hole with a perpendicular back. The tree is placed against this perpendicular face and the soil packed firmly against the roots. These must be straight down and the tree at the same depth that it was in the nursery. A supply of seedlings is carried in a canvas bag and the roots must be kept moist at all times. All trees on hand on a project are kept in beds with earth closely packed around the roots. They are removed from these "heel-in-beds" as required by the planting crews. The normal size of a planting crew consists of fifteen planters, one staker-packer, and a crew foreman. The crew moves across country roughly line-abreast with each man keeping six feet from his neighbour. Each man plants trees as he goes at approximately six foot spacing, varying, of course, for stumps, logs, rock, and other obstacles. The man on each end of the line usually has a line of stakes to guide him. Strips are normally laid out to run up and down hill. Coming down hill the planters travel backwards, more or less. With an ideal six-foot spacing, 1,210 trees per acre would be planted. The actual figure is usually around 900 to 1,000, but may go as low as 600 depending on the type of ground, debris, and amount of natural reproduction already on the ground. The staker-packer is responsible for keeping guide lines ahead of the crew and keeping the planters supplied with trees. The crew foreman must supervise the planting one hundred per cent of his time, and he is constantly checking the work of each planter.

BC's forests had to be regrown, and they required assistance if this was to happen in a satisfactory manner. Therefore, protection from wildfire needed to be increased. Distant areas could no longer just be left to burn, as every tree lost

The launch *Forest Surveyor* at the Forest Service Marine Station in 1964. The Forest Service maintained a large fleet of vessels of various sizes and types. NA-22175.

to fire was one that could not be harvested. Critical to Victoria's control of the new programs was increased and better communication and the ability for field officers to reach more distant areas of the province. The Forest Service's already large marine fleet expanded, with more launches traversing the coast, and lake and river boats became everyday tools in most ranger districts. There was also a gradual increase in the use of fixed-wing aircraft and, later, helicopters, until they became the central means of access to the more distant points. As the road network expanded, often built by the Forest Service, car and truck access to remote areas became possible. Radio use, while employed in the past, expanded and revolutionized administration procedures by providing a relatively quick and convenient way to convey urgent messages.

Dirk Trigg: Tenures Officer, Kamloops

The ranger district office was in Horsefly and with a satellite office in Likely. A very interesting district. It had the huge Quesnel Lake and Horsefly Lake, and there was a lot of work that was done on boats, boat access up the lakes. Huge volumes in a lot of the stands, and it was back in the days when the districts were different. They were very autonomous. Before the roads were paved all

The Horsefly Quesnel Lake road, through immature timber, 1953. NA-14973.

the way out to the district offices, it took a lot longer to go in and out so you didn't travel into Williams Lake. You might go every second or third week to get groceries or whatever. And so you became part of the community and the district office actually.... It would probably be frowned on today, but it became the centre of a lot of the friends of the foresters. The loggers, whatever, would show up at the district office after hours on a Friday night, and there was a bit of a community get-together. Before going home for dinner or going home to see the wife or whatever, you had a bit of a community get-together. A lot of things revolved around the district office, the old ranger station. So you're kind of the centre of the community. Everybody kind of knew who you were. They were smaller communities and you were the centre of the community, and they knew we were the guys to go to ... there wasn't a special protection branch. They knew the guys in the district office not only gave out the cutting permits but also, if you had a fire, these were the guys who were going to go out and fight the fire.

Expansion of the radio network began almost immediately on the close of World War II. In 1945 Victoria radio traffic to the districts went up by

Servicing radio equipment in the Victoria repair shop, 1957. NA-17026.

70 per cent. Many of the types of units required could not be purchased commercially, and so the building and experimenting with radios in the Victoria office increased. The use of radio by every branch of the Forest Service expanded to such an extent that in 1952, when the Engineering Services Branch was formed, the radio section was moved from Protection to Engineering. In total in 1952, 617 radio stations were operational and 80 portables were available for use. The Forest Service also gradually shifted from AM to FM. By the time Orchard retired in 1958, the radio network had more than doubled, with 1216 radio units in operation.

Finding the Forest: Surveys and Inventory

How much forest existed, what type of forest, how old was it, and how fast did it grow? The Forest Service had only sketchy answers to these questions, but answering them was critical to successful forest management and especially to the success of sustained yield. The Forest Service did not know, in many areas, even how much timber was available for commercial harvesting under a sustained-yield program.

Survey vehicles and equipment at the Patricia Bay airport in 1953, being readied for the season's field work. NA-13824.

Inventories had been in progress since the Forest Branch was first organized in 1912. The then chief forester, H.R. MacMillan, sent out survey parties to regions of the province in an effort to at least gather an idea of the trees he was expected to manage. This was frequently the first task of the new forestry graduates coming to work for the Forest Branch. P.Z. Caverhill's first report (he later became chief forester between 1920 and 1935) was a survey between Hazelton and Burns Lake in 1912. The report was only 15 pages long, but it did make recommendations regarding fire protection and noted the locations of merchantable timber as well as other features, such as agricultural possibilities.

New inventories began immediately after the branch was established. Chief Forester MacMillan knew inventorying the forest was critical if the government was to actually control harvesting of the forest resource. As the Fulton Royal Commission report pointed out in 1910, estimates of British Columbia's timber resource had been hugely inflated. The commissioner noted "forest statistics have been, in the immediate past, the wildest guesswork, and even recent revision … is based on very little information."[5]

Although the reconnaissance work was much reduced by the Great War, inventory and cruising work at specific locations continued, despite the drain of manpower and financial resources caused by the war. The very specific inventory work required to allocate timber licences was quite different from the general reconnaissance, which tended initially to be less-than-thorough surveys of a region. F.W. Beards' report on the Adams River Valley, about 160,000 acres (65,000 hectares) in extent, was only 12 pages plus two maps.

Three reconnaissance reports were completed in 1912, but survey work was started on at least 10 projects, and this expanded to 12 additional finished reports in 1913. In all, a total of 5,616,000 acres (2,273,000 ha) were examined in 1912, with a further 12,308,000 acres (4,981,000 ha) in 1913.[6] These somewhat superficial reports, even by 1920 standards, began to provide a basis for understanding the province's forests. Work had progressed far enough that a preliminary forest map was published in the 1913 department annual report. Although the map still only differentiated heavy timber, lighter timber and timberless land, it was a start at cataloguing the forests of BC.

By 1920 the reconnaissance work resumed on an expanded scale, with significant improvements in the standard of reporting and the extent of the areas surveyed. In 1932 D.L. McMullan described the work of the surveys division. The Forest Branch generally, at this time, undertook two types of surveys: extensive reconnaissance surveys and strip surveys. The extensive reconnaissance survey was considered a preliminary survey and provided a rough guide for when a more intensive survey could be undertaken. The reconnaissance crews

Taking the age of a Lodgepole Pine using an increment borer, 1956.
NA-16025.

usually consisted of two men – a cruiser and a compassman – and the area covered could be quite extensive. The strip surveys were much more intensive – smaller areas with larger crews. The strip surveys were run by several crews. Usually, the strips were one mile apart, and a crew could run a two- to three-mile strip each day, recording a wide range of data for transposition onto maps. Topography, timber (both mature and immature) and land classification were the fundamental objectives of the strip surveys.[7]

The first comprehensive review of all of British Columbia's timber resource was undertaken in cooperation with the BC Forest Branch by the Canada Commission of Conservation. H.N. Whitford and R.D. Craig published *Forests of British Columbia* in 1918. In 1927 inventory took on a new life, and reports were published covering most of the province between 19?5 and 1938. In 1935 F.D. Mulholland, as the forester in charge of the surveys division, also began preparation of a revised and more comprehensive study of British Columbia's forests. Mulholland's report, published in 1937, can be considered the first comprehensive review of BC's forests based on comparatively thorough surveys, although it still only included about 13 per cent of the estimated potential area. It covered a broad range of information and provided a resource of considerable importance at that point in time.

In the period following the 1945 Sloan Royal Commission, as British Columbia moved toward sustained-yield management and as the industry grew by leaps and bounds, the existing information was clearly inadequate. Immediate action was required to correct the situation. The forester in charge of the forest economics division, which now included the surveys section, was F.S. McKinnon, and he immediately sought to answer the questions of how much forest was actually available and how much of it could be harvested on a sustained-yield basis. In 1945 he sent out the first survey party in four years and, using a six-man crew, they began inventorying the central west coast of Vancouver Island. In 1946 he was able to place two crews in the field, both working on Vancouver Island.

Survey crew on Toba inlet, 1957. NA-16831.

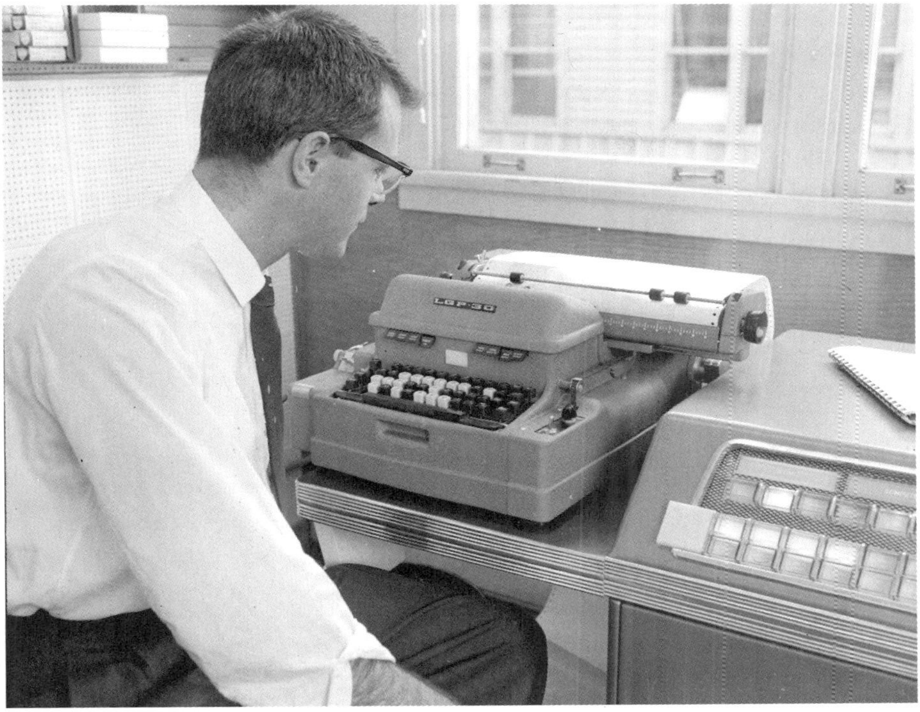

Using the LGP 30 computer. NA-19566.

In 1947 there still were not enough personnel, especially with experience, but the Forest Service managed to field four crews, and the first data was being compiled and mapped for the Clayoquot region. McKinnon estimated in 1948 that the Vancouver Island inventory would be completed in the next year. He started to move crews into other areas of the province, including Smith Inlet, the Kitimat and Douglas Channel area, the Upper Fraser area and Princeton. In 1949 five crews totaling 60 men were in the field. The importance of this work and the scale of the growing activity were recognized on January 1, 1951, when a new surveys and inventory division was formed under the direction of H.M. Pogue.

A federal-provincial forest agreement in 1951 resulted in significant federal involvement. The federal government would now pay 50 per cent of forest inventory costs. This allowed the division to field 278 staff in 1952. Cars, trucks, boats, airplanes and helicopters were all used to get crews to their survey areas as quickly and conveniently as possible. A radio network was set up using 70 portable radios to keep crews in touch with the various field headquarters. Air photos began to be used to monitor depletions from the forest base.

In 1957 a major inventory of the whole province was published, but still the work continued. Management of the forest required detailed inventories and regular monitoring of the depletions to the forest base. The initial inventories now had to be checked and rechecked and rechecked again.

Federal funding enabled the surveys and inventory division to expand into new technologies. In 1956 punch-cards were introduced for mechanical compilation of data, and by 1959 the proven value of computers resulted in the acquisition of an LGP-30 computer. First manufactured in 1956 in California, the LGP-30 was among the first desk computers. In 1959 when inventory division received 54 requests for statistical summaries of specific areas of the province, it was by using the computer that they were able to meet this demand. By 1957 inventory division was getting air photos of sufficient quality that they could now estimate forest volumes using photo-mensuration techniques. In 1958 the division provided, for the first time, complete forest map coverage at one inch to two miles, which required 149 sheets. A gigantic map using all the sheets was created and placed on display at the Pacific National Exhibition in Vancouver. It was also in the 1950s that the BC Forest Service hired their first woman professional forester. Francis Robinson (nee Wetton) began working in the inventory division in 1955. She was soon followed by two women graduates from the Sopron School Edith Andódy and Elisabeth Juhász, who also joined the inventory division.

As their work was refined, the surveys and inventory division began spending more time studying the wide range of factors that affected forest regrowth, as well as what factors other than logging would cause forest depletion, in an effort to refine data even further. The more accurate the data, the better the chief forester could judge what the harvest should be in each managed area.

Inventory remained a key aspect of how the Forest Service controlled the harvest rate. As with all divisions and branches, their role continued to grow. Not only were they responsible for determining the timber potentially available over the long term, but the division began to investigate a broader range of inventory elements. By 1978 they had completed 31 environmental protection mapping projects in Public Sustained Yield Units. Growth and yield surveys, and even the number of study plots, continued to expand. Further new technologies also allowed improved services, one important step being the introduction of a new computerized mapping system, the Interactive Graphics Design System, which was implemented in 1978.

Almost all forests in BC are publicly owned (about 94 per cent), and the responsibility for managing and maintaining inventories and standards has always rested with the government, although in some instances the holders of

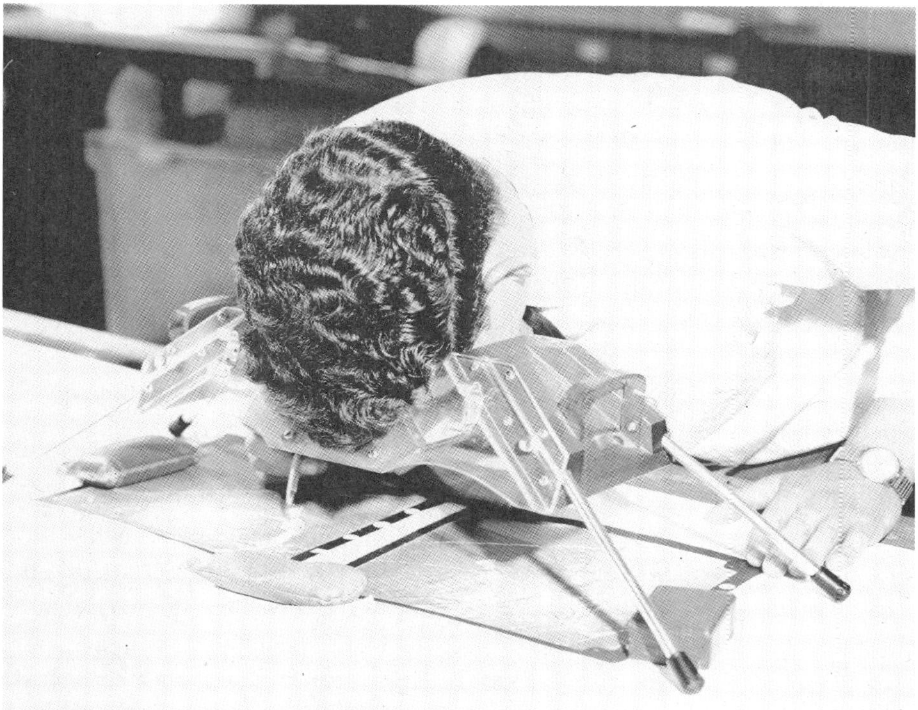

Using table stereoscope and air photos, 1964. NA-22123.

tree farm licences have also played a key role. Given the extent of BC forests, forest inventories are a critical necessity supporting good forest management and land use planning.

Forest Service programs and projects became ever more sophisticated as time moved on. In 1992 and 1993 inventory division staff developed a variable density yield prediction model to help with determining likely stand yields. When the carbon budget model was introduced, the Forest Inventory group, developed by the Canadian Forest Service in the late 1980s as a means to mitigate the effects of climate change, played a critical role in assessing the British Columbia situation. Forest inventory continues to play a crucial role in assessing the extent of British Columbia's forests and in assessing not only industry harvesting levels but the numerous factors associated with understanding and properly managing our environment. In recent years maintaining accurate inventories has been made especially challenging by events such as the Mountain Pine Beetle infestation in the interior, which has changed the nature of the landscape over a large part of British Columbia.

Shaping the Forest

In order to manage the forest through sustained yield, difficult decisions had to be made quickly. Within months of the 1947 legislation, the demand for timber far exceeded the supply, and the Forest Service needed the government to stand firmly behind the sustained-yield policy if there was to be any hope of successful implementation. Many companies also quickly realized that the forest management licences (which became known as tree farm licences during the 1950s) provided them with the most control over timber supply. Under the forest management licence, the lumberman took on a much greater responsibility for roads and for reforestation, and while each area logged required a cutting permit, the timber was generally managed by the company albeit with Forest Service supervision. But the main thing was that the timber supply was assured; no bidding at auction occurred and therefore there was no opportunity for anyone else to sneak in and purloin that timber. In the Public Working Circles there was always the risk that the applicant could lose the bid and, therefore, its timber supply, despite assurances by the Forest Service that the established loggers would not be put out of business. It was also intended that the Public Working Circles would provide the many small loggers with a living, even though the government preferred the larger operators, who they perceived as better able to work with the sustained-yield program.

H-frame jammer loading logs onto a truck on Forest Management Licence No 8, operated by Boundary Falls sawmills, Windfall Creek, 1951. NA-12485.

A portable sawmill on the Zymagotitz River, near Terrace, 1949. Hundreds of small sawmills sprang into existence and constantly demanded timber. NA-12104.

The provincial government and the BC Forest Service plunged forward, intent on making British Columbia timber management a model for other jurisdictions to follow. Orchard announced in 1952 that, in his opinion, British Columbia's forest policy was "the most progressive … in North America."[8] The policy as Minister Kenney expressed it was that

> Forest management simply seeks to keep every forest acre under growing forest-crop; protect the growing and mature timber from fire, insects and disease; limit the annual harvest to a volume not in excess of the average growth and constantly increase the quality and quantity of the harvest. For the first time in our history the logger can grow timber – becomes a farmer….[9]

Despite the Forest Service's long acceptance of the idea of sustained yield, the detailed information needed to establish a managed forest did not exist. Yet decisions had to be made, and delay could spell disaster as lumbermen clamoured for timber to supply their mills. Most of the big coastal mills had sufficient timber in existing tenures to run for a time, but they wanted more and they would not wait for long. It was a different story in the interior. New operators sprang up daily, insistently demanding access, while the longer-established operators tried to tie up as much timber as they could in secure tenures. One company even asked for a secure supply covering the whole northern half of the province. Already established operators continued to receive timber sales as requested and were considered as practical by the local ranger and the district forester. In the interior most claims for forest management licences were rejected (only a few relatively small forest management licences were granted), while most of the interior forest was allocated to Public Working Circles. This meant the Forest Service could exert more control over the cut-and-run small operators who dotted the landscape. In contrast, on the coast more timber was locked into large forest management licences. In this way the Forest Service was able to bring company-owned lands into a sustained-yield management program. The established coastal Public Working Circles did provide some opportunity for smaller loggers.

The first of the forest management licences was issued in 1948 on the north coast, to Columbia Cellulose on the promise of a new pulp and paper mill near Prince Rupert. The second, which included scattered areas of Vancouver Island and the south coast, was issued to the long-established Canadian Western Lumber Company in 1949 and similarly based on the promise of the construction of a new pulp and paper mill, near Campbell River. It was also on the south coast that in 1949 the Forest Service began to define the first Public Working Circle: the Sayward Public Working Circle, forming part of the Sayward Provincial Forest. Under the direction of Davis Carey, the Sayward Public Working Circle was defined by 1950 and a second, Sechelt, was well underway. Many unresolved issues, however, began to delay progress. As McKinnon noted to Orchard, one difficulty in creating the Sechelt Working Circle was that no decision had been made on which forest management licence applications were going to be approved. E.W. Bassett, Vancouver district forester, also noted his concern that allowable cuts should not simply be set up on a cruiser's recommendation. He argued that as there were few areas in the Sechelt Public Working Circle that allowed for long-term sales, it would be more appropriate to issue timber sales without an allowable cut attached. The allowable cut should be for the whole Working Circle, and each sale should be at a set cut. The frequent overruns could then be charged against the whole Working Circle's allowable cut. Eventually, a version of this method is what proved to be most practical. Similar concerns delayed and caused problems in many of the proposed Public Working Circles.

Glitches in communication and process also needed to be worked out, and sometimes communication between Victoria and the district offices was not as smooth as it should have been. When instructed to implement the Sayward and Sechelt Public Working Circles' allowable cuts, Bassett very pointedly wrote:

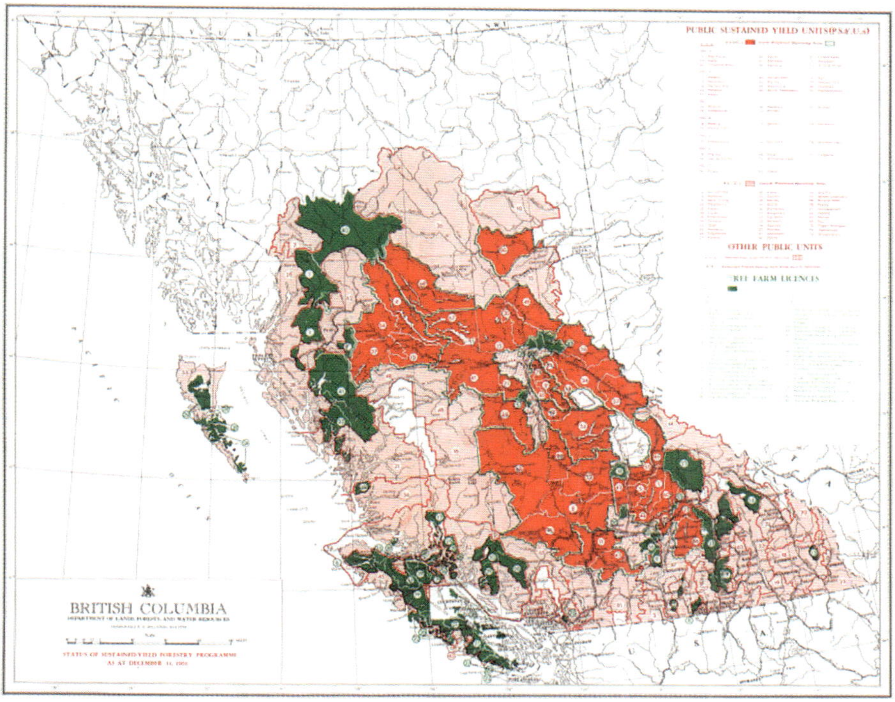

Map showing the location of Forest Management Licences and Public Working Circles.
RBCM research files.

"We have apparently in this office never been advised of the finally approved figure of allowable cuts on either of these forests.... Since we are now to advise all interested parties formally in respect to this cutting budget, I feel we should have definite advice of the figures finally decided upon in respect to each Forest."[10]

In July of 1950 Carey and three graduate foresters were working on planning Public Working Circles. At the request of Assistant Chief Forester R.C. St Clair, McKinnon advised Orchard that for 1951 they were considering expanding in a number of locations. In 1951 Carey and his assistants Mason, Parlow and Groner expected to finish Powell River, Salmon Arm and Chilliwack.[11] In order to regulate the Public Working Circles, Victoria senior staff decided that the allowable cut would be monitored by cut control ledgers maintained by the districts, with copies in Victoria, and with the district's ranger responsible for the area being cut. These were in place for Sayward, Sechelt and Chilliwack by February 1951. In addition, a record of cutting priorities was to be developed for each Working Circle, which would be used as part of the decision-making process as to whether or not a timber sale would be issued.

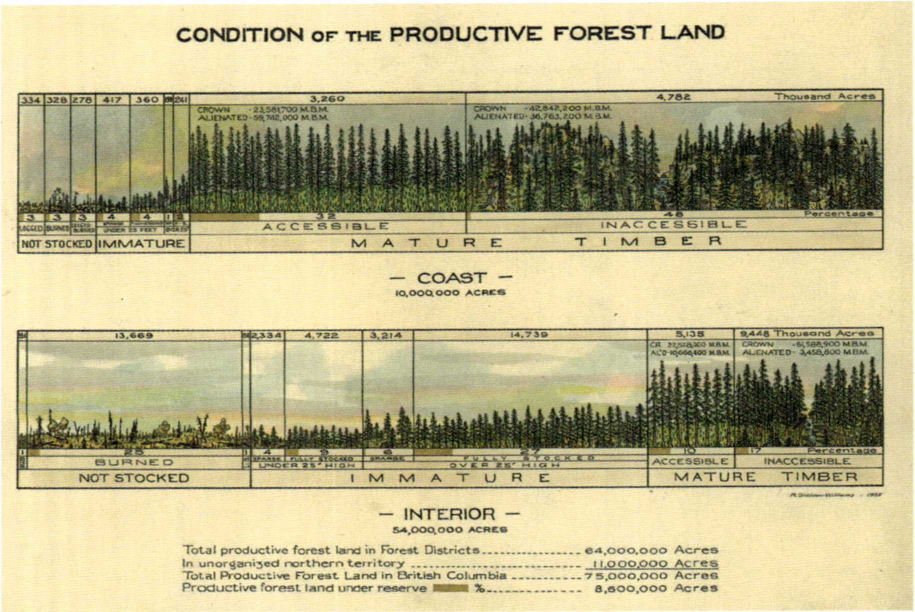

F.D. Mulholland created an inventory of the provincial forests in 1936. These images show the information, images and map of one such forest: Upper Arrow. CM 18621 Y.

E.W. Bassett, 1948. As with many who eventually ended up in senior management, Bassett had a varied career. Before being appointed district forester in Vancouver, he was in charge of operations in Victoria, and following his stint in Vancouver he returned to Victoria as assistant chief forester for operations. Later, he became deputy minister of Lands. NA-13399

Industry demands for timber meant that events had to move much faster than initially planned, however. By May 1951 Sayward, Sechelt, Chilliwack and Crooked River were all considered to be under sustained-yield regulation, and five additional areas were being reviewed for inclusion: Barkley Sound, Salmon Arm, Spallumcheen, McGillivray Lake and Seymour. Preliminary work had also been undertaken to delineate additional coastal Public Working Circles. As well, by the end of 1951 there were six signed forest management licences in place, a further eight under reservation and 117 active requests for forest management licences. The allowable cut on 11 coastal Public Working Circles had been calculated, and preliminary work was undertaken on 15 interior Public Working Circles.[12] By 1959, although much of the far northern part of the province was still not operated as managed forests, there were 36 forest management licences (by this time called tree farm licences) and 66 Public Working Circles, gradually being renamed Public Sustained Yield Units.

Turning a Public Working Circle into a managed forest was no easy task. Loggers were busily cutting timber in almost all the areas prior to the boundaries being set and certainly prior to the allowed annual cut being established. As each

Public Working Circle was created, ledgers were started in order to record the timber, including the areas licensed for logging, and the depletion rate. When an area was proposed for management on a sustained-yield basis, the Forest Service compiled a list of those already logging the area and met with them to explain what was happening and how they should seek to be recognized as working in that sustained-yield unit. Victoria staff, district officers and rangers all worked with the industry to reach agreement and to implement sustained-yield practices. The Forest Service was determined to make this process as fair as possible and to ensure that no one during the initial phase was left without timber. Orchard noted that at one meeting, to be held in Prince George, he expected more Forest Service staff would be present than interested parties but that he thought this was fitting, as he wanted to emphasize the importance the Forest Service placed on such meetings.

One example of the Forest Service working collaboratively with districts and loggers was in the central interior. An early Public Working Circle named Naver, located between Quesnel and Prince George, was established in 1951. In this case, the extension of the Pacific Great Eastern Railway from Quesnel to Prince George had created significant interest in the area's timber, but fortunately, comparatively little actual logging had taken place. In order to establish a managed

Logged area in the Naver Public Sustained Yield Unit, 1968. NA-24520.

Loaded logging truck at Okanagan Lake, 1972. NA-28039.

area before the timber rush, District Forester Lorne Swannell recommended to Victoria that they establish a Public Working Circle. With no inventory for the area, they had to act quickly. By mid 1951 Prince George staff determined that 14 loggers and sawmillers had timber sales in the area but were cutting slightly less than the established annual allowable cut (22,500 foot board measure). There was some concern because Northern Spruce and Eagle Lake sawmills both had forest management licence applications covering parts of the area and no decision had been made as to their disposition. McKinnon praised the Prince George forest district staff for their work in a letter to Orchard: "This is a very worthy effort on the part of the Prince George staff and is to be highly commended."[13] Most management efforts were not that straightforward. The Okanagan Public Working Circle, established in 1953, had 83 different loggers and sawmillers identified as having rights in the area. Frequently, the many timber licences or overlapping claims in more settled areas exceeded the purposed allowable cuts, and this sometimes resulted in difficult negotiations.

In 1952 the working plans division was established, initially under the direction of W.G. Hughes and then that of M. Pogue, to speed the process of creating new management forest units. In 1965 the working plans division, with most of the province's forests now under sustained-yield management, was merged into the forest inventory division[14], and in 1966 the division was eliminated. Many of its duties were transferred to the management division, while the responsibility for calculating allowable cuts was shifted to the inventory division.

Pumps and Lookouts

The BC Forest Service made a substantial contribution to detecting and fighting fire during the 1940s and 1950s. Regular liaison with the US Forest Service opened up new avenues for the BC Forest Service to explore, and important new developments resulted. The mechanical men at the Fraser River Forest Service Marine Station were very inventive. Testing and use of pumps had come under the Marine Station's activities, and Superintendent Macdonald and Forman Bennett involved themselves in developing several pieces of equipment.

The Lookout

R.G. McKee designed and proposed a new prefabricated lookout building in the early 1940s, with two experimental lookouts completed 1942 and 1943. He drew on many standard features, such as lightning arresters,

Prefab lookout under construction at Livingstone Mountain. Private Collection.

but incorporated a radical departure in construction by using plywood in prefabricated panels. The result was a substantial savings in time and money. In a report, McKee described in detail the benefits from such construction and also how the process should proceed, from the initial prefabrication process right through to completion of the building. By the end of 1947 seven of the new designs were in service and the Forest Service was rapidly expanding its lookout system using the design. Within four years, 50 so-designed lookouts were active. The Forest Service continued to use this design, and modified versions of it, for the next several decades.

Fire Finders and Pumps

J.G. MacDonald and M.K. Bennett liked to tinker, and with more than the boats they built and maintained. Together they developed several improvements to devices that assisted firefighting.

Lookoutman at the Terrace lookout using the fire finder, as improved by Bennett and MacDonald, 1972. NA-27739.

MacDonald was very familiar with the Forest Branch/Service, having joined 1912. He worked as a ranger and district forester and held other positions before becoming superintendent of the Marine Station in 1946. The Forest Service had long realized that if they could get a crew on the ground at the start of a fire, they could frequently control it and keep it from spreading. Studies in the United States had long shown the best methods to attack a fire, and *The Western Fire Fighter's Manual*, prepared by the Western Forestry and Conservation Association, was used as the basis for firefighting in British Columbia. Critical to a rapid attack was being able to accurately locate a fire, and to this end Bennett and MacDonald developed a new and more accurate fire finder. Using the fire finder, the lookout man could accurately report the direction of a fire from the tower, and if two lookouts could view the same fire, they could precisely pinpoint its location.

Testing Bennett-MacDonald pumps at the Forest Service Marine Depot. NA-08909.

This device, combined with the panoramic photographs, provided extremely accurate location information.

Once the fire was located, getting men to the site was critical, but they needed equipment to fight the fire. MacDonald had been working on a pump for some time, and the two men decided to create a new and lighter pump. What they sought to design was a pump that could be carried in a backpack. When the first test models were finished, 20 test units were put into service in 1947, with four sent to each district. The pump, along with tools, two hours' worth of gas and 100 feet of hose plus six feet of suction hose, fit comfortably into a carrying pack weighing only 40.5 pounds (18 kg). The cost of construction, excluding the hose, was $105.31. E.W. Bassett, in charge of operations (protection division), noted "much credit demonstrated in the production of this unit."[15] MacDonald and Bennett continued to tinker and also developed a new fuel moisture stick gauge to assist in predicting fire danger.

H. Bell checking fuel moisture sticks and indicator, Langford, 1963. NA-21805.

The End of an Era: A New Government

C.D. Orchard found the later 1950s to be a period of disappointment and frustration. Following his cordial relationship with the coalition government and particularly with Edward Kenney, Orchard viewed the new Social Credit government of W.A.C. Bennett with suspicion, and they soon had doubts about him. Bennett's agenda was development, not restraint, which proved a mixed blessing for forestry. For a while, the resources available for the BC Forest Service's programs grew, and on the surface it seemed all was improving. But a long, tough road lay ahead.

After 10 years of intense effort, J.S. Stokes, forester in charge of management, confessed to the second Sloan Commission that the Forest Service still did not have final working plans for all of the 64 Public Working Circles theoretically being managed. The cut allocation was still largely based on guesswork. In many instances, Commissioner Chief Justice Gordon Sloan thought that the Forest Service was erring on the side of caution. It later proved, though, that even cautious estimates were optimistic, and considerable difficulty was emerging in regard to overcutting. While the Bennett government sought ways to increase economic development and increase harvesting, the Forest Service continued to try and bring the harvest under control and create a management system that they believed would bring about stable, forest-based communities and achieve long-term forest harvesting.

A Crew at the Vancouver District office working up field notes, 1956. NA-15878.

Chapter 5
The Williston Years, 1956–1972: A New Agenda and Style

For several years after the 1952 election, Orchard carried on as deputy minister and chief forester. Despite the wariness the Social Credit government showed towards him, he would have been difficult to dislodge, especially given the influence of Gordon Sloan and the thoroughness with which Orchard undertook to implement the Sloan Commission's recommendations. When Orchard retired in September 1958, Ray Williston had been the Minister of Lands and Forests for over two and half years. When he took office in 1956, Williston was faced with a disheartened staff and a lingering controversy over the way timber was being allocated in the Tree Farm Licences, a controversy that had led to the previous minister, Robert Sommers, spending a little over two years in jail for accepting bribe in allocating a Forest Management Licence.

Ray Williston, minister of Forests, 1966. NA-22402.

Confronted with enormous problems and the massive report from Gordon Sloan (printed in 1957), Williston initially continued to follow the policies implemented by Orchard without significant change. On the completion of his second Royal Commission, Sloan became forest advisor to the government and so continued to enjoy considerable influence. It was only after Orchard retired, the year after the commission's report, that Williston began to make significant policy changes. First, he separated the positions of deputy minister and chief forester, appointing R.G. McKee as deputy minister and Finn McKinnon as chief forester – both having been long-serving members of the BC Forest Branch and Service. Williston advised the more senior, McKinnon, that he was viewed as Orchard's hatchet man and that he must prove to Williston that he was something more. Subsequently, Williston began to look more closely at policy and to place his mark upon both the Forest Service and the forest industry.

Adjusting the Agenda

Williston remained committed to sustained yield as outlined by the Sloan commissions – at least, within limits. But the glow of the sustained yield policy was wearing off, and flaws were evident in the program. Also, the Social Credit agenda of industrial development meant the forest had to be reevaluated. The task was massive. Without enough resources and still faced with inaccurate,

Pulp Mill at Gold River, 1972. I-07201.

inadequate data and an entrenched industry, it was near impossible. Enormous efforts were directed at inventorying and cruising the forest as accurately as possible, especially the specific timber sales. But the immense variety of land and forest made it extremely difficult to collect the necessary information. Forested areas of critical concern were placed into managed areas as quickly as they could be – where, hopefully, the cut could be regulated to meet the long-term needs of industry and therefore provide economic stability. On the coast, where much of the timber was in tree farm licences, progress in controlling the harvest was seemingly being made, especially when large companies supported the program. Companies such as Alaska Pine & Cellulose and the Tahsis Company invested heavily in reforestation and set cutting plans that seemed to make sustained yield attainable. Not all of the industry was as impressed. The smaller lumbermen, excluded from "the Favoured Forty" argued that "the concept of the TFL may be sound, but the basis of ownership is odious."[1] For the Forest Service, however, it was a practical option that seemed to work.

Despite steady increases in staff from 1945 to 1972, the task of reevaluating the forest was more than could be possibly accomplished in so short a time. It seemed a miracle that as rigorous a regulation of the provincial cut occurred as did, in fact, happen. But it remained difficult, and one assistant forester complained: "I have your note of tribulation that you want all Working Circles under sustained regulation. So I offer my apologies and for the next ten months I'll grow arms like an octopus and do everything at once!"[2]

Three interior public working circles (sustained-yield units) seemed especially to be in serious trouble, and they were declared emergency areas. These were all in the Kamloops forest district (Lac La Hache, Williams Lake and Nicola). In Williams Lake the problem had been compounded by Orchard when he decided to divide the large Cariboo East Public Working Circle into two smaller Public Working Circles: Williams Lake and Quesnel Lake. Although it made sense from a management standpoint, this division led to serious difficulties for industry. In the heavily overcut Williams Lake area of the divided (now called) Public Sustained Yield Unit, a duel emerged between the minister and the Forest Service, who clashed with the initial 83 loggers, to bring the cut under control and to encourage the loggers to move into the undercut Quesnel Lake region. Williston, in part sparked by the issues in these overcut Working Circles and under pressure from the logging industry, began to change timber sale regulations in order to provide more protection for existing operators. If an operator/applicant was prepared to meet the bidding price, they would secure any applications they had initiated for timber that was approved by the department. It should also be noted that by this time, only someone who held

Jacobson Brothers Sawmill at Williams Lake, 1979. I-04926.

timber rights could apply for a new timber sale in a Sustained Yield Unit. In difficult areas like Williams Lake the minister not only set policy, but he also frequently made the operating decisions, as the lumbermen tended to bypass the Forest Service and appeal directly to the minister to solve their problems. Williston generally supported his staff by repeatedly advising the lumbermen and the community of Williams Lake that the area was overcut and that the Forest Service was going to continue to hold the line on allocation. But it was not unusual for Williston to make public announcements and then let the Forest Service determine how the announced policy would be accomplished.

Warning signs were being raised in other districts as well. In a study of Naver Unit 2 area, a memo to the district forester in 1962 noted that "at face value the situation is downright alarming. We can hardly say we are practicing sustained yield in the light of such a short cutting period.... The end of the first cutting cycle is in sight.... The Natural Question is – what went wrong? The answer – nothing!"[3] The memo proceeds to mention a variety of reasons why over harvesting might have occurred and indicates several possible solutions. The problem of excessive harvesting was correctable, but it was indicative of growing concerns, especially with a provincial government determined to sell as much timber as possible.

The Forest Service engineering program continued to open new forest areas, and by the time Williston left office in 1972, only the areas northeast of Dawson Creek and around Telegraph Creek and Cassiar were not under sustained-yield regulation. Once the engineering division had constructed an access road and an area was opened for timber applications, it was not long before lumbermen moved in to begin harvesting. And as the number of regulated areas increased, so did the difficulties rangers and district foresters faced.

Meanwhile, the inventory division was still only able to supply baseline information for some of the areas now supposedly under regulation. Between 1961 and 1972, 77 sustained-yield units and special sale areas had been re-surveyed, mapped and comprehensively inventoried; 16 still needed to be completed.

The Recreational Forest

The Forest Service's parks and recreation division was a unique organization. The Parks Branch was formed in 1939 when the Forest Ministry was given responsibility for parks. Parks operated under several different divisions, including forest economics, until it was set up as its own division in 1948. Despite differences and the occasional conflict over timber, the Forest Service

C.D. Orchard and H.R. MacMillan at the opening of Manning Park, 1950. NA-10137.

Picnic area at Little Qualicum Falls Park, 1952.
NA-13181.

was extremely proud of their involvement in establishing park and recreation areas throughout British Columbia. The small group tasked with maintenance and the establishment of new parks worked hard. When the parks and recreation division was moved to a separate department in 1957[4], along with the Game Department and Tourism, the Forest Service and C.D. Orchard felt that their support of parks had been betrayed. As a sign of their commitment, the Forest Service did not simply drop all aspects of recreation, but continued to establish recreation sites throughout the province.

In the 1940s parks fell under the forest economics division, a group of miscellaneous activities that included research, reforestation and surveys and inventory. The Forest Branch in 1945 was administering 53 separate parks that included 16 Class A, 4 Class B and 28 Class C parks, along with five parks administered under separate acts. Gordon Sloan's recommendation, in his first report, of a separate parks section in the BC Forest Service was quickly acted upon. Staff threw themselves into the work, although efforts to hire additional staff initially met with little success. The parks and recreation division moved from a holding pattern between 1941 and 1945 into a period of rapid expansion and became a hive of activity. New equipment was acquired, and by the end of 1948 there were 58 parks operated by the BC Forest Service. Earlier experiments in prefabricating signs and tables had proven valuable, and a new shop was built in Langford. Initial efforts at planning park use began at Mount Seymour, and reconnaissance for new parks and research into park use and trends were actively pursued.

Although considerable additional funding was available in the late 1940s, so much work needed to be done that the most urgent projects had to be highlighted. In 1948 the parks section focused on five separate areas: Little Qualicum Falls Park, Peace Arch Park, Manning Park, Wells Gray Park and building the Langford workshop. Although most resources went toward these parks, some effort was directed toward other parks as well – for example, a sign for the new park at Cathedral Grove and, invariably, some work at Mount Seymour. Much energy was spent on fixing trails at Mount Seymour in 1948, and similarly in Tweedsmuir Park. As well, some major projects were initiated in 1948 in several of the larger parks. In Mount Seymour and Manning parks, preliminary work began on new facilities. The new Pine Woods Lodge in Manning Park was one of the highlights when it was officially opened on June 3, 1950. Park improvements were to continue on at this pace for the next decade until the parks section moved to its new department.

View of Battle Mountains, Wells Grey Park, 1952. NA-11849.

The evaluation of potential new sites also expanded, with 75 potential sites examined and several existing parks re-examined to determine if the existing boundaries were actually the best. The division began the process of creating master plans for each forest district; this was spurred on by the numerous applications around Kamloops for fishing camps and lakefront properties. Although only a start was made on a master plan, about 40 different lakes in the Kamloops area were examined for their recreation potential.

The Parks Division began to spread the workload, appointing regional recreation officers. The first was appointed to Kamloops forest district in 1950, and other districts soon followed. The regional recreation officers spent much of their time working on the smaller sites. W.H. Hepper reported in 1956 from Nelson that most of his time between mid June and September was spent supervising youth crews and park attendants, while a little time was also found to search for new recreational reserves to serve as camp or picnic sites. Of the eight campsites under his purview, the one at Wasa Lake was the most heavily used, and despite improvements, demand was greater than its capacity.

By the last year that Parks Division was part of the BC Forest Service, 1956, the number of parks grew to 112, and this did not include the 1024 recreational reserves, called UREP reserves (meaning for the Use, Recreation and Enjoyment of the Public), that protected the picnic areas and campsites scattered around British Columbia. Park attendance had now reached nearly 2,000,000 yearly visitors, with Mount Seymour receiving the heaviest use, followed by Miracle Beach.

One of the main goals of the W.A.C. Bennett government and Ray Williston was the development of an interior pulp industry. To achieve this goal, two new policies were implemented. The first was called third band timber utilization, and the second was the inauguration of pulpwood harvesting area agreements.

Third band sales were generally based on utilization of smaller wood from the harvesting areas and were awarded based on improved sawmill technology, especially the introduction of barkers, small log gang mills and chippers. This, Williston hoped, would encourage regional sawmills to supply chips to the pulp mills that he was also encouraging be built. Third band timber was not allocated in a casual fashion. Detailed studies were developed of the companies that indicated they were interested and could process third band timber. The Forest Service required that they meet specific criteria, including having installed or being in the

Logs in Babine Lake at the Babine Sawmill, 1949. I-26322.

process of installing certain equipment. In the Nechako Public Sustained Yield Unit, for instance, only three sawmillers of 38 met these criteria. Each of the mills interested in third band timber very carefully informed Williston and Deputy Minister John Stokes of their capacity to meet Forest Service requirements. For example, Fraser Lake Sawmills wrote in November 1969: "we have converted all our plants to close utilization standards...." [5]The complexity of who should receive an allocation of third band timber is illustrated by the final decision in this Public Sustained Yield Unit. The four major operators were all provided with substantial allotments, but in addition, nine smaller operators who were tied directly to the larger mills were given small quantities, and three additional operators were also given third band timber even though they were not otherwise operating in the Public Sustained Yield Unit. Rustad Brothers, a Prince George sawmiller, was given an allotment, as there was not enough available third band timber in the Crooked River Sustained Yield Unit to meet their needs. Williston's efforts to secure a suitable chip supply from the sawmills worked.

But it was not enough to inaugurate a new pulp industry. In order to provide the security required by capital investors, Williston also offered areas to the pulp companies on which they could harvest pulp wood to meet a mill's chip requirements, should the regional sawmills not be able to supply adequate fibre. These were designated as Pulpwood Harvesting Areas. Only in a very few areas where the several licences were issued did actual log harvesting take place to supplement chip supply. The rest of the interior pulp mills rapidly became dependent on chips supplied by regional sawmills. The pulpwood harvesting areas also provided among the first occasions that open public forums were held to discuss each specific licence. In an effort to overcome the negative backlash from the Sommers affair, Williston began to conduct public reviews on selected tree farm licence applications as well.

By the late 1950s and early 1960s the Forest Service's argument and recommendation to disallow all future applications for tree farm licences in the interior was accepted by the minster, and the majority of the interior forests

Prince George pulp-and-paper mill, 1965. NA-22565.

Ray Williston signing Pulp Harvesting Agreement No. 1, 1962. NA-21630.

were subsequently managed in Public Sustained Yield Units. The clamour from many of the small operators was growing too loud to ignore, and it was hoped these operating areas would help solve the difficulty.

Getting the Message Out

Orchard sought to convince industry and special interest groups, including the BC Forest Service, that the great experiment of sustained yield would work. Forest Minister Kenney's job, along with the newly created public relations and education division, was to reach all British Columbians. Orchard became an inveterate conference attender, something that he could not have done without support from his senior staff. Sometimes he was accompanied by the minister, sometimes by other staff, and sometimes other staff led the way. When Orchard did attend, he was invariably seated at the head table, whether it was a United Kingdom Timber delegation or a meeting of the Interior Lumber Manufacturers Association. His reputation

C.D. Orchard speaking at the 6th Commonwealth Forestry conference in 1952. NA-13142.

preceded him, as did the significance of what the Forest Service was seeking to do. When he attended the 5th British Empire Forestry Conference in London in 1948, he was the only Canadian speaker. In 1949 he became president of The Association of British Columbia Foresters, replacing Fred Mulholland, a former BC Forest Branch employee and, at the time, chief forester for Canadian Western Lumber Company Limited. In the same year that the BC Timber delegation left for a tour of Norway and Sweden, Orchard was a member of the group along with F.S. McKinnon, who succeeded Orchard as the province's chief forester in 1958.

When the Northern Interior Lumberman's Association met in Prince George in 1947, District Forester R.G. McKee (later deputy minister) attended and spoke at a luncheon where he cautioned the loggers and mill owners about the problem and hazard of debris accumulation – an opening salvo in the Forest Service's campaign to reduce fire hazard and burn slash. During the Canadian Society of Forest Engineers' tour of Vancouver Island, Harold MacWilliams, the head of the reforestation division, met the challenge and arranged for the group of 55 foresters to visit various locations to view a variety of reforestation projects, including an examination of

Vancouver Island tour by the Royal Forest Society, 1962. NA-21550.

beetle problems and a visit to the Campbell River Nursery. In 1950 Orchard was replaced as president of the Association of British Columbia Foresters (formed in 1947 by legislation) by George Allen of UBC. The vice-president was Harold MacWilliams, and on the executive was E.W. Bassett, the district forester for Vancouver, and Lorne Swannell, the district forester for Fort George. Orchard moved on to be elected president of the Western Forestry and Conservation Association for 1952.

The BC Forest Service's involvement in these organizations continued throughout the 1950s. This was clearly a vehicle they could use to convince the public and the world that indeed the BC Forest Service was on the right track. When delegates of the Sixth British Commonwealth Forestry Conference toured British Columbia in 1952, more than 50 members of the BC Forest Service ensured the visitors had an excellent time and reinforced the view that "British Columbia's growing 'experiment' in forest management and integration" received its deserved recognition as a "scheme under which private industry and governmental services can co-operate with mutual benefit."[6] The small but active research division also regularly joined in these activities. For instance, when the Northwest Forest

Genetics Association met in 1957 at the Duncan Forest Nursery, they were welcomed by Assistant Chief Forester McKinnon, but the main participant was research division's Dr Alan Orr-Ewing. He discussed his research into forest genetics and explained the "plus tree" program.[7]

The Forest Service's new public information and education division, formed in 1945 under director Eric Druce, immediately began to launch full-scale campaigns. By the end of the second year, he and his staff were attending public meetings, creating exhibits and getting the message out in a variety of ways. Films related to forestry, for example, were shown to 371 different audiences totalling an estimated 32,633 persons. The three most popular films were *Exploring Tweedsmuir Park*, *Land of Timber* and *Garibaldi Park*. The division's work grew quickly. They created an exhibit for the 1947 Pacific National Exhibition in Vancouver and, by 1949, had their own darkroom and photography program.

Druce and his staff worked at improving program delivery at the forest district level. Education had long been a part of a ranger's work – giving talks, attending meetings and speaking to school classes. Druce's division now assisted with this work.

As the 1950s progressed, the operation became more varied and more organized. In 1952, for example, 6300 feet (1900 metres) of motion picture

Barbara Davies and John Andre shooting film footage in Victoria, 1959.
NA-17630.

film was produced. The division created a vast resource of information relating to the forest and forestry in British Columbia: radio spots, TV spots, a trailer for commercial theatres, exhibits, roadside signs, reports and publications, school talks, and lectures throughout the province to a variety of groups. Information was also spread through the Forest Service library, and additional assistance was given to district and ranger offices. All of this was directed at furthering the education of both the public and industry, not just about the BC Forest Service but about what the industry and the forests meant to British Columbians.

The Burning Forest

Deputy Minister McKee told the Western Forestry Conference in 1961 that fire control had to be first in British Columbia's forestry objectives. As Charles Cowan of the Washington Forest Protection Association wrote, "The Enemy is Fire."[8] Protecting the existing forest from hazards remained a central challenge for the Forest Service, and wildfire was the number one enemy. McKee pointed out in 1965 that "without protection (from fire) there can be

The Forest Service were very serious about fire prevention and this dramatic sign was placed along the highway in Manning Park, 1957. I-28674.

The Grove Fire near Prince George in 1961 burned 40,000 acres (16,000 ha) of forest land.
NA-20781.

no sustained-yield forestry."[9] The BC Forest Service were not the only ones preoccupied with fire; industry was as well. If a wildfire was allowed to rage unchecked through a tree farm licence or a Public Sustained Yield Unit, pulp mills and sawmills would shut down soon after.

The 1960s was a time of considerable change in the way the Forest Service controlled wildfire. The disastrous fire seasons of 1958, 1960 and 1961 and the continued problems of finding and fighting fire increasingly drew on resources and expertise. Although improvements in detection were made, such as increased use of aircraft, the main innovations of this time were in fighting fires, and it was to fight fire that the greatest organizational changes were made.

While the use of aerial attack and heavy equipment was emphasized more and more, tried and true firefighting methods also remained critical, and constant reminders were required to ensure that the basic principles did not get lost. Harry Forse, in charge of the Forest Service's protection division in 1961, announced: "If you want to put out a forest fire get men on it and don't ignore the warnings of the hydro-thermograph, fuel-moisture indicators and the meteorologist."[10]

Avenger dropping bentonite, 1960. NA-20409.

The use of aircraft for spotting and controlling fire became the strategic solution of the 1960s. The experiments of the late 1950s demonstrated such good results that industry was also starting to invest, and five of the larger coastal companies bought four Martin Mars cargo transports for conversion into water bombers. These became the "New God of Rain."

Fixed-wing aircraft dominated the new approach to firefighting, although helicopters began to exert greater influence as the decade progressed. In 1960 the BC Forest Service contracted for six Beaver aircraft and one Super Cub, but by 1970 the Forest Service had under contract 11 Avengers, each carrying 500 gallons; 3 Cansos, each carrying 800 gallons; and experiments were made using 3 Douglas A-26s, each carrying 900 gallons. Each group included one or more bird-dog aircraft used to coordinate the aerial resources. The fleets were stationed at Prince George, Kamloops and Vancouver. A variety of small craft were also used as fire patrol. Helicopters were routinely contracted in each district as required, although Vancouver maintained one on standby due to a shortage of machines in the area. In addition, the contracting companies had extra aircraft ready and waiting to take up wildfire work in demanding years.

Avengers formed the backbone of the contract fleet during the 1960s, with Cansos and occasionally the Martin Mars providing assistance. The first of the Cansos was converted by Calgary-based Cal-Air Service Ltd in 1962, although they were not extensively used until 1965. The company purchased and converted six Cansos and based three in British Columbia – at Patricia Bay, near Victoria – and three in Alberta. The Cansos particularly intrigued the Vancouver forest district, as they found the Avengers not especially satisfactory for working on the coast, although they were a very welcome addition in the interior districts. In 1961 the Vancouver district protection officer noted: "Despite some enthusiastic reports by company personnel we have still to see water bombers operate effectively on a fire on the Coast. This is borne out by the majority of rangers and field staff who have had the opportunity to observe water bombers under actual fire conditions." In 1966 the Vancouver district changed its tune upon the introduction of the Cansos, and the protection officer was very impressed by how effective they were on the Hood Fire.

The Forest Service had also learned that bird-dog aircraft played a critical role by keeping communications open between ground and air and pinpointing the

Canso, Avenger and A-26 water bombers at Kamloops, 1970. NA-25438.

Avenger water bomber pilot John Cline, 1962. NA-21131.

water bomber targets. Even though the Forest Service only contracted 12 Avengers for standby in 1962, additional aircraft were generally available. For example, Skyway Air Services, the main contractor for the Avengers, had 15 in readiness. The availability of the water bombers required some operational reorganization. In the Nelson forest district a squadron of three Avengers, leased from Skyway, was stationed at Cranbrook, along with a bird-dog aircraft leased from Kootenay Airways Limited and flown by Ranger Bailey. The squadron attacked 17 fires between July 11 and August 2 and kept all but two of the fires to less than an acre. The ground crews were ferried to the hard-to-reach fires by a leased Okanagan Helicopter. The Forest Service also experimented with new and larger aircraft in the late 1960s. Although they looked at the Canadair twin-engine CL-215, it was the Douglas A-26 that caught their attention, and in 1970 three were under contract.

Bob Perchie: On the Use of Air Tankers

Generally, what you're going to do is box the fire in. In some instances, you might also coat it so it will keep the crowning from happening, particularly in the hotter part of the year. Once you're finished, then you go back and wait for the next. As I mentioned earlier, generally in the operational season there are two of you covering off one tanker.

So you fly one week with a group, and then the next week you are off doing other things. Often what I would be doing.... I always tried to go to a number of fires to take a look on the ground to see how effective it was from the air.

The other advantage I had was: part of my responsibility was to ensure consistency in applications. I was involved a lot with the initial attack crews, how they were trained. So I was able to go out into a district and without getting too involved look at the incidents and see if, in fact, not only were the air tankers effective but how they went about establishing their guard, to insure consistency in the approach. It was a great life.

Martin Mars water bomber at Patricia Bay near Victoria, 1960. NA-19809.

Flying near fires was a dangerous business, and potential aircraft disaster was a recognized enemy. In 1967, for example, although it was a higher than usual accident rate, five pilots and crew died. While none of the dead were permanent employees of the BC Forest Service, it was a chilling reminder of the hazards of firefighting from the air. In the first accident, David Hill of White Rock died when his Avenger crashed near Fruitvale. Two more were lost on July 17, when Alex Davidson and Paddy Moore's Canso crashed while fighting a fire on Skirt Mountain near Victoria. On August 20 two more died: Tim Ross and Jack Fornell, while dropping supplies to a fire crew near Nakusp. An updraft tangled a chute in the plane's tail assembly, and the plane smashed into a hillside while the ground fire crew could only watch in horror.

The Forest Service was also beginning to look more closely at helicopter use, and not just for transportation of men and supplies. Okanagan Helicopters was demonstrating to the Forest Service the potential of bucket-equipped helicopters. The company's Nelson manager, Jim Grady, had designed a bucket from a 45-gallon (170-litre) fuel drum, and eventually the 45-gallon monsoon bucket became the standard equipment for helicopter drops. Although their

Helicopter attacking the Mil Fire, 1960. NA-20262.

Tool cache at the Courtenay ranger station, 1948. NA-09179.

use for such work was sporadic, helicopters were occasionally used because of their precision drops. By the early 1970s additional trials were being conducted, and interest in the helicopter as an air tanker grew. The Forest Service began comparing the various buckets available and their capacities, including the Rainmaker from the United States and a design that the Forest Service developed called the Fielder Mark I, which had a 120-gallon (450-litre) capacity.

In terms of other equipment innovations, the basic backpack hand pump was being improved by making it lighter and easier to manipulate. As early as 1963 Wajax, BC's leading supplier of pumps, was offering a polyethylene hand pump nearly four pounds (2 kg) lighter than their earlier galvanized model. Wajax's (Watson-Jacks of Montreal) Mark 2 pump had been generally adopted by 1955, and these older models remained in use as long as they were effective. Wajax continued to improve pump performance and dominated supply during the 1960s and 1970s, although some of the now-ancient Bennett-MacDonald pumps (see page 168) were still being used in the Vancouver district in the early 1960s.

In addition to equipment changes, the Forest Service expanded its detection and suppression expertise and procedures throughout this period. Efforts included the development of a backpack hose layer, an improved drip torch for lighting back-fires, a small field propane stove, a new hand-held fuel moisture stick indicator and, in 1969, a fire simulator for the indoor training on fire suppression tactics. Some organization changes occurred as well. Staff were repositioned to meet problems in communications and wildfire management. Although 8- to 10-man suppression crews continued to be placed at various locations, use of the two- and three-man initial attack crews increased. Often carried in by helicopter, they established an initial base and could frequently extinguish small fires before they spread. The team also initiated fire control on the bigger blazes until the larger crew could arrive. In the Vancouver district, although manpower remained almost the same, the suppression crews were reduced by closing one camp and removing one man from the Nanaimo and Alberni camps. Instead, six initial attack crews were scattered around the district. The initial attack crew carried a 60-pound (27 kg) pack with enough food for four days, plus a pulaski and lady shovel. If this equipment did not suffice, the teams were quickly reinforced with auxiliary equipment that included a heavy-duty pump and a light backpack pump, plus hand pumps.

Much of this change was induced by severe fire seasons in the late 1950s and early '60s. 1960 saw an tremendous struggle against wildfire. On July 13 a massive lighting storm raged across southern British Columbia, and 599 new fires were quickly reported. Campfire and burning permits were suspended in Kamloops and Nelson forest districts and most of the Vancouver forest district, and numerous general closures were also placed on watersheds and other areas. In total, 4700 men were employed fighting the fires, along with 200 tractors. Water bombing became the major weapon, and every available aircraft was pressed into service. Assistance was even obtained from the RCAF, with two Piaseki Helicopters and two Otters. One pilot was killed near Kelowna, and a Flying Fortress water bomber crashed on the Kamloops airport, necessitating a special United States Air Force crew to be flown in from Washington State to clear the wreckage. In all, 4124 fires had been fought and extinguished by October of 1960. The two largest fires were the Dean fire near Merritt, covering 13,000 acres (5261 hectares), and the Gal fire 25 miles (40 km) from Kamloops, covering 10,000 acres (4047 hectares). It was also an unusual year in that three individuals were charged with lighting a forest fire and were successfully prosecuted for arson.

Dean Fire, stopped by a new fireguard, Kamloops District, 1960. NA-20221.

Despite being a terrible season, the new procedures that had been implemented and continued to be improved through the 1960s had proven themselves. In 1958, with approximately the same number of fires, over two million acres (809,371 hectares) had been burned by wildfire. But in 1960, with better access and more reliance on aviation, the fire loss had been kept to 285,820 acres (115,667 hectares). Accessibility, though, remained a serious problem that was only gradually solved when another disastrous fire season emerged in 1961 – the season with the highest incidence of fires, with well over one million acres (404,685 hectares) scorched by wildfire, often in inaccessible areas. The improved practices were further demonstrated in 1967, which saw the worst fire season on record. Despite the record 3,216 fires that year, the area burned was kept to only 255,000 acres (103,194 hectares) of blackened forest

Grazing Lands

Range management had been part of the BC Forest Service's mandate for decades. Much of British Columbia's rangeland is also forest land, and integrated planning

Douglas Lake Cattle Company cattle near Minnie Lake, 1960. NA-20080.

was essential for its successful management. The emphasis that the BC Forest Service placed on range management varied year by year and decade by decade.

Range management first came under the BC Forest Branch in 1919, and in 1920 Thomas P. Mackenzie was appointed commissioner of grazing. Mackenzie was a great believer in stock and range associations and worked tirelessly to help organize associations to help in the management of the various ranges. Much of the fieldwork was undertaken by rangers, but interest varied considerably. By 1921, 22 range associations were active, although Mackenzie noted that many ranchers were apathetic about range management. In order to encourage more active participation, Mackenzie saw to it that Range Improvement Fund applications would only be accepted from the range associations. This funding was made available for a variety of projects, ranging from constructing drift fences to fixing mud holes or building corrals on Crown land.

Mackenzie also began a series of range studies in an effort to understand the various problems and to convince cattlemen of the best way to manage the ranges. Experimental seeding plots were established, but by 1926 these initial efforts proved to be a failure. Efforts were also directed at controlling insect infestations, such as grasshopper outbreaks. One major concern were the wild horses, which caused considerable range damage and even broke fences, as aggressive stallions would set mares free to supplement their herds. During the 1920s an aggressive program was implemented to cull the wild horse populations.

By 1935 Mackenzie joined the federal research program at Kamloops, and G.P. Melrose, along with his other duties, was appointed commissioner of grazing. Although the programs already underway were continued, Melrose had little time for or interest in pursuing new initiatives.

During the 1940s the grazing division was part of Operations, and its director reported to the forester in charge. In 1949 grazing once again became its own division and was placed under the direction of W.C. Pendray[11], who reported directly to the assistant chief forester. When the Forest Service was reorganized in 1952 into two branches – Technical, and Planning and Operations – the grazing division director once again reported to the assistant chief forester for Operations. This remained the case until 1974, at which time grazing was shifted into the Resource Management Branch and renamed the Range Division in 1975. The Range Branch was formed in 1978, and its headquarters are presently located in Kamloops.

Division staff relied extensively on rangers and their staffs for field assistance. The rangers and assistant rangers received some training in range

Cattle in timber near the road to Face Lake, 1960. Many range areas were on forested lands. NA-20091.

management – at least those who had attended ranger school – but proper management required an agrologist to coordinate the work. Range planning therefore remained somewhat erratic until the Forest Service placed agrologists on district staff. The first placement occurred in Kamloops, the largest grazing district, and then in Nelson in 1947, followed by Prince George in 1956. Range technicians were also eventually placed in some ranger offices. As with every aspect of Forest Service responsibility during the 1950s, the workload increased significantly. New range reconnaissance studies were made, general improvements were made to ranges (often on the recommendation of a local stock association), and staffing levels increased. Most rangers, though, seemed to appreciate the efforts of the district agrologists.

When Jim Milroy, later division director, pointed out in his 1956 report on the Nelson district that there were growing conflicts between ranchers and hunters, several rangers immediately realized they were going to have to put more emphasis into reconciling these diverging interests. Milroy thought that for the first time, he had completed, in 1956, a year of comparative "humdrum mediocrity," but he was also quick to point out he had no desire for one of the

Jim Milroy, 1977.
NA-33893.

"violent outbursts" of previous years.[12] Although Milroy thought it a quiet year, it was a busy year. Considerable agitation on the part of the ranchers emerged, led by the Waldo Stockmen's Association, in opposition to Milroy's efforts to change the opening date from mid April to May 1 to protect the ranges. The opposition was so great that his decision had to be reversed, and he hoped the ranges would not suffer too much as a consequence. He also faced agitation from chambers of commerce and tourist boards complaining about the hazard of cattle on the roads. Unfortunately, there was little he could do, as these cattle were generally found to be coming off private rangeland rather than Crown range. Meanwhile, the ranchers were blaming the rod-and-gun clubs and the Game Department for the large population of deer and elk now resident on some ranges, claiming the animals were damaging the range. These complainants conveniently forgot that it was, in fact, the ranchers themselves who were doing their best to keep hunters off the ranges. Milroy also investigated biological control, but using beetles to control Hypericum (St John's Wort) did not seem to be working as well as hoped. Attention was then turned to controlling weeds by others means. Range seeding was carried out on seven stock ranges, seven drift fences were built or improved, three cattle guards were installed, and one mud hole was dealt with. Although 1956 may have seemed humdrum to Jim Milroy, it proved to be a very busy year.

Perhaps the issue the grazing division would face every year can be summed up from Milroy's somewhat philosophical introduction to his 1957 report to his District Forester:

> Grazing Administration, we believe, has come a little distance in the past decade but comparatively speaking it is a long way from keeping pace with scientific jet age developments. Some of the policies of the Service, the attitudes of the public in general and of our grazing clientele in particular which frustrate progress in rangeland management and improvement.... Moves designed for the betterment of the resource or for the industry as a whole involve certain basic investments that must be paid in advance. Revenues at the present are not such as would justify sizable government expenditure to increase the crop of forage from Crown Ranges. So we fudge along as best we can. We build a drift fence here and a water development there as funds allow.

He continues on presenting his view of what should and should not be done and ended by saying: "We are united as one, however, in wishing the Service … a genuinely prosperous 1958 with at least some of the improvements forthcoming towards which all have been struggling so valiantly in the years gone by."[13]

In 1932 the grazing division headquarters moved to Kamloops in an effort to place it closer to main cattle raising areas. It remained there until 1945, when it was again shifted to Victoria. Staff continued to work in the districts with agrologists, eventually in Kamloops, Nelson, Prince George, Dawson Creek and Smithers. When the Cariboo forest district was reformed in 1972, it included as much rangeland as Kamloops. In 1978 grazing again received a new mandate, under the *New Range Act*, with the addition of 15 new range technicians. A process of assessment and reorganization began. The Range Management Branch director now reported to the Assistant Deputy Minister for Timber Range and Recreation. In 1988 range management became a program of the Integrated Resource Management Program. In 1992 range became a separate branch, now in the forestry division, with its director reporting to the chief forester.

Tom Wallace, assistant agrologist at Kamloops, crossing a snowbank near Pukeashun Mountain, 1953. NA-14085.

The grazing office in Kamloops, 1950. Left to right: Bruce Neighbor, Jim Maxwell, J. Pinder-Moss (behind filing cabinets) and Dorothy Hopgood. NA-10601.

By 1997 range was considered a section of the Forest Practices Branch, where it remained until 2005. In 2006 the Range Branch was recreated and its headquarters moved to Kamloops, where the current director now resides.

Gordon Comeau: Retired Range Technician

But what happened before the agrologist was to be more of a service: to monitor, make sure they're not overgrazing, and meeting all the criteria, to dealing with it more on a scientific level. The agrologist played a critical role in getting us to look at range in a different way, using his skill and what he's learned in his university days and that kind of stuff.

Planning became a more integral part, where we would draft plans with the rancher as to how they would manage their cattle, setting rotation systems for the range, where they'll be certain periods of time. Rest rotations – saying every year you will rest one portion of your range to allow it really to recover. So they started to develop all of these range use plans. And we would have to do that.

In the later years, it even got to the point where we would go out and identify the different plant communities in the different parts of the range and write up what the range plant communities were made of – picking up all the different species of the grasses and the different vegetation that was out there. And from that, they could determine, really, the condition of range. So it became from more of a service to the rancher to more of a planning tool to help the rancher maintain a viable range, still a service to the rancher.

The rancher didn't always agree that when you told him his cattle had to come off, it was time for the cows to come off. "Now, I've been grazing it this way for so many years," he would argue. But we based our planning on when the stubble height gets four inches on certain species of grass, like Kentucky Blue Grass. That's when the cattle will come off.

We also specified by species height when the cows could actually go out. And if they wanted to turn out early or something like that, they would have to come

H.K. DeBeck, at the Henstridge Prairie, Adams Flateau, analysing forage using an aerial photograph to map range types, 1953. NA-14147.

and apply. We would go out and look to see what the condition of the range was. Before, there was just a specific date that says you can turn out on May 15, and you turn them back in on September 15. Those dates still applied. But sometimes you had a beautiful spring, you could turn out early. And it was based basically on when the Pinegrass, for example, started to fold over, which was at six inches. Then the range was ready for cattle to be moved on to it.

So, it became more scientific and more of a planning method of administration than it was a local person service.

Glen Heyes: Retired Agrologist, Merritt

You know, a lot of the ranching industry, you get from just a mom-and-pop family operation up to the big – what would it be? – like, the industrial scale, like a timber industry. And it's.... I would say my sense of that is it comes down to judgment and how you interact with people. You can't have different laws for different people and, you know, the larger ranches would emphasize that. You know, we want to be treated the same, or the same privileges and so on. It came down to kind of judging the situation and doing what's sensible, and I think we always had the latitude to do that.

And the context that I'm thinking of, one of the things that needed to be done is there had to be a range use plan. And this is kind of an operational plan that describes how the range is going to be used. And there's the larger ranches – this is certainly an easier task for them to generate this plan. And the Forest Service in, I guess, the senior levels – they want to promote the ranchers doing the plans, I think for two reasons really. One is they thought it was a cost-cutting measure. And, you know, have ranchers do it themselves rather than pay the Forest Service to have to do it. And the other aspect was: when someone generates a plan themselves, they feel there's more ownership in it and, you know, they'll definitely know it and it'll be a better plan in that respect, and possibly a labour plan entirely, because he may know his range better and so on.

What we found, was that for some people this was a real difficult chore for them, and it just was far more efficient for us to write the plan for them, even though the law was saying something else and the direction was something else. What I found worked really well is you have a discussion with them, find out what they wanted to do. We knew what we wanted them to do. You'd get a meeting of the minds through a discussion, whereas for some it's a pain in the butt for them, right? And things like mapping – they can't generate mapping and that kind of stuff, so it was kind of a hybrid and you'd probably…. I guess it's fair to say you put more attention in where more attention was needed.

And that was probably maybe a little bit different than the timber industry, where you would have professional foresters or consultants and so on generating these things. The ranching community couldn't afford consultants, and there wasn't really a ranching consultant industry big enough to support that anyways. But it's far easier to be uniform, I guess I would say, with the timber industry than it was with the ranching industry, but it worked out. In my view it worked out and worked fairly well. I mean, it always boiled back to: what's the objective of this whole exercise? And the objective really is to get the land looked after. The rancher has a limited amount of resources, and if you siphon them off doing other stuff, he's going to have less resources to apply to doing that job, that main objective, which is to try and have that land looked after.

David Borth: Branch Director

We've got ecologists. We've got people that are really focused on ecology and range ecology specifically. We've got what we call a practices specialist, and he's focused on riparian and upland grazing areas and what's the best way to make use of those areas, what's the level of harvesting, the level of grazing you should do. We've got invasive plant specialists.

Delivery of the range program is done… I'll back up. You mentioned I got my specialists here in the branch. The people out in the districts are more general. I mean, they deal with the tenures and they deal with all the things I just talked about, but they're more general. They don't report through to me; they report through to their district managers and the regional manager and, ultimately, my boss. That's where it comes across. I have a branch of specialists.

But our branch is actually totally operationally oriented. I mentioned invasive plants already. That's 20 to 30 per cent of my staff of 25. We are leading the charge on Crown land to keep plants from invading the forests and grasslands, you know, invasive plants. And we have people on the ground doing it.

Then over on another area is ecosystem restoration. In our sense, in the range program that is managing the areas that have historically been maintained through fire – open forest and grasslands historically maintained through fire. So we use some of the same tools. We use slashing and timber harvesting and burning to maintain these open areas. In the Cariboo we're actually getting Mother Nature to do it this year. I've got a manager that looks

John Chisholm's sheep near Elko, 1949. NA-10568.

after the ecosystem restoration for these natural disturbance types that are open forest and maintained by fire.

Then we work directly with communities, local communities and their restoration committees to do that work. Backing up, we do the same thing. There's invasive plant committees. People in communities are very much interested in getting engaged in managing invasive plants and maintaining the forest attributes in the southern interior that are maintained by fire. So we're working with community groups.

The third major area we got programmed and the branch is directly engaged is working with the livestock industry, like the BC Cattlemen's Association deliver on fencing programs. Mountain Pine Beetle came in and impacted trees falling on fences. Fires then burned through all those fences. So we're building, rebuilding, maintaining, really, a public infrastructure around fences that facilitate good management out on the range. So we're doing that. And we've even recently partnered with the Ministry of Transportation to look at fences along highways.

Myriad Activities

By the 1960s the British Columbia Forest Service was able to begin assessing its management of the forest. Forest management was always something of a dilemma. The Forest Service sought to understand and preserve a forest that had to be cut in order to generate revenue for the greater good of the province, and it acted under direction from elected representatives who needed help in understanding the nuances of managing a forest in perpetuity. Despite changes in government, industry policy and practice, the Forest Service, under Williston as minister, continued in the direction established in the 1950s, while striving to simply keep ahead of the changing economic and social situation surrounding forest management. Improvements were made in all areas, with better reforestation practices, improved access to difficult regions, better communication to the public and to each other, and a better understanding of the forest. In addition, that somewhat detached group responsible for administering the cattle ranges, the Grazing Branch, continued to fulfill their responsibilities.

Vancouver Forest District general office in the Marine Building, Vancouver, 1956. NA-15867

New Foresters

The Sopron School of Forestry in Sopron, Hungary, had strongly supported the 1956 Hungarian Revolution and opposed the invasion of Hungary by Russian troops, and a decision had to be made for students to flee or stay. On November 4, 1956, 450 students (about half of the school's total student body) along with about 50 professors fled into Austria. Of these, 220 were forestry students. One month later, on December 4, Canadian immigration officials informed the forestry school refugees that they had been accepted as a group to Canada and that the University of British Columbia would maintain the school until the last class graduated. The Powell River Company financed the endeavour. Initially, the group landed in St John's, Newfoundland, and only finally reached British Columbia on January 24. After a few side trips, including a stay in Abbotsford and then at Powell River, the group headed for Vancouver and finally reached UBC on September 10, 1957.

It was a difficult transition, as most of the Hungarian students spoke very little English and being kept together as a group meant there was insufficient emphasis placed on learning the language. Other aspects also came as a shock, such as learning that they would individually have to pay university fees and that they were expected to find summer work to meet this requirement. The training program in Hungary was quite different from that provided and even required in Canada. The intensive forestry practised in Europe was not happening in British Columbia, and while new courses and instructors were added in an attempt to learn about local conditions, much of the program and the examinations were allowed to remain unchanged.

The first class of 27 graduated in 1958 and the final class in 1961. Most of the group requested and were granted Canadian citizenship in 1962. In all, 141 of the Sopron students graduated at UBC, and many remained in British Columbia – although as the years passed, more and more spread throughout Canada and even to other countries, with most working in forestry-related occupations. That 202 students had come to Canada and only 141 graduated is an indication of how difficult the transition was. A number of graduates were employed by the BC Forest Service and included some of the Service's first women foresters, such as Edith Andódy. In 1968 she completed her thesis for the Association of British Columbia Foresters. The contribution of the Sopron foresters to the advancement of forestry in British Columbia is undeniable. Throughout the 1960s, '70s and '80s the Sopron graduates worked in the private and public sectors and academia. Most foresters and technicians employed in BC during that period either worked with, alongside or for a Sopron graduate at some point in their careers, and many have fond recollections of this unique episode in BC forest history.

Lyle Resh: Retired Agrologist

In 1972 the Cariboo forest region at Williams Lake was formed from parts of the Kamloops and Prince George forest regions. I returned as a forest agrologist at Williams Lake under Ed Smith I/C, with a staff of three forest agrologists. I was assigned the Chilcotin area to work with Rangers Tom Walker (Tatla Lake) and Emil Bouchard (Alexis Creek).

Ten years since I was last in the Chilcotin, it appeared things were softening up towards the Forest Service and government. As Jim White, agrologist from Kamloops, spent a lot of his time with the ranchers during the grazing season, ranger staff were always overrun by fires and other forestry matters and spent little time in the field in grazing matters during the summer. Tom Walker had assigned Herb Spence, a young assistant ranger, to help in the grazing work. It took time to have some trust built up between the ranchers and ourselves. We spent long hours on horseback, walking, vehicle or helicopter use, explaining our side and listening to range users, which led to a better working relationship. Now and again the relationship would slide backwards, but that was to be expected.

In later years, when Lodgepole Pine trees were clear-cut logged, with this came loss of barriers and numerous roads which let cattle drift into adjacent range units, causing mixing of herds which they were not permitted for. Clauses were added to the cutting permits, and the logging companies had to build drift fences and install cattleguards to help solve some of the problems created by this type of logging. It also created forage for the cattle to graze by allowing light in on the suppressed pine grass, which then began to grow abundantly. But livestock did not like this type of grass as much as others, like Bluebunch Wheatgrass. They had to be more or less enclosed in these Pinegrass sites for better utilization.

Livestock association meetings were not as confrontational as the past, except if a touchy subject was presented by us, such as a later livestock turnout in the spring. We had noted for some time that the turnout dates in the spring were too early; consequently, we launched a range readiness program. We discussed it with the associations that we would give the ranchers five years to build up their hay base and then move turnout dates from May 1 to May 15, or later in some areas. If some years the range was ready sooner, we could adjust. It was a two-way street, and at times the rancher would only turn out part of the herd on the opening date. On a lot of open range areas very little climax grass species as Bluebunch Wheatgrass survived. There now were a lot of pussy toes and other weeds, which would take over. We were usually told Bluebunch Wheatgrass never existed when they first started ranching. It was not true. It depended how long the ranch was in operation and their use of Crown range.

The sites that had little grazing by livestock were also looked at, such as islands in the middle of a lake or further back in the trees off of open range sites. Fenced exclosures of a quarter acre or so were set up to keep the cattle out of the area and to monitor how long it would take to come back to a climax condition. Small wire cages were also used to measure grazed verses ungrazed for that year. Our objective was to have 50 per cent utilization by domestic stock, which left further grazing by wildlife and plant vigor for the following year. Some open range dry sites took over 20 years to come back to climax species. With readiness checks and in the end, livestock turnout dates were moved to May 15 or later.

Each branch and division worked toward the same goal: protecting and maintaining a forest that could be economically harvested. The Forest Service's focus on this aspect made it seem that employees had little appreciation for the other values that were slowly emerging in regard to preservation and recreation – that is, creating a multiple-use forest for the enjoyment and use of many, not simply a revenue generator for the government's treasury. Forest Service personnel were keenly aware of these other values and issues, but as an organization they were required to implement activities based on a legal framework. An example of the Forest Service's changing interests is when one of the district forest staff tried, in 1973, to interest the Parks Branch in creating a provincial park at Tzenzaicut Lake in the Nazko (about 100 km west of Quesnel). When the Parks Branch decided against this, the forester who had made the suggestion then convinced the Forest Service's senior management that this area should be withdrawn from potential harvest and set aside as a recreation site.[14] Such activities were not unusual within the Forest Service and included discussions around opening logging roads for recreation purposes, as well as the careful selection of special areas for preservation.

The Forest Service's legislated mandate, however, was to provide access for harvesting, secure revenue for the government and ensure a forest remained sustainable for future harvesting. Investigation of scarification and tree planting suggested ways the forest could be regenerated. In 1962 $100,000 was

Site preparation for reforestation using scarification. NA-14101.

committed to a program of scarification in various interior spruce and true fir forests. Scarification's purpose was to expose the mineral soil that would assist in spruce seed germination. Well aware that natural seeding was not going to replenish the forest, the nursery program continued to expand. In 1964 about 15 million trees were planted, and in 1965 this was expected to increase to 25 million. By 1970 over 33 million trees were being planted annually. The plus tree program was also expanded, and seed orchards with cone production based on grafts from these "super trees" increased. It was also in 1968 that the Forest Service first hired female tree planters. These women planted 145,000 two-year-old spruce seedlings in the Prince Rupert forest district under the direction of Dave Wallinger of the reforestation division. Experiments to increase seedling survival rates were also implemented. The most successful was developed by Weldwood forester Norm Pelton, who worked with the Forest Service to create a system of packing seedling roots in mud and then encapsulating them. Using a dibble, the rate of planting increased markedly and survival rates jumped from 70 per cent to 90 per cent.

One of the largest projects undertaken by the BC Forest Service during this time was clearing the region behind the new W.A.C. Bennett Dam, which was flooded to create Williston Lake. Using six D-9 Caterpillars, the Forest Service cleared about three acres (1.2 hectares) at a rate of an hour per machine. Approximately 1500 cubic feet of timber was salvaged per acre. It was an immense project, as the lake was to cover about 640 square miles (1660 square km). After the first year the Forest Service worked from float camps, and it was expected that the toppled trees would float to the surface where they could be salvaged. Although considerable wood was salvaged through this process, much more work needs to be done. One of the more notable experiments was the use of LeTourneau Tree Crushers. The tree crushers would go over land that

Floating logging camp on Williston Lake, 1968. NA-24251.

held non-merchantable timber and knock down every tree in front of huge steel wheels that would then partially grind them up. Although there is some question as to how successful they were, two of the machines were used for several years. One of these is on display at Mackenzie, BC.

While many tend to think of the Forest Service in terms of the branches, divisions and sections that make policy or put people out on the land, departments like accounting and personnel also made significant contributions to the organization. Personnel, for example, arranged district forester meetings, ranger meetings, zone meetings and protection meetings. The division also administered the comings and goings of staff. Turnover in 1970 was 9.2 per cent amongst the 906 permanent staff, down significantly from the 12.3 per cent of 1969. In most years Prince Rupert suffered the greatest disruption, with a turnover of 16.2 per cent of all staff. Research also remained active and continued to investigate a wide range of topics intended to both understand the forest and to improve its value. Studies such as delineating the ecology of interior forests contrasted with more prosaic but still valuable studies, like frost hardiness in Douglas-fir seedlings.

Dave Caughlin: Assistant Ranger

Areas are logged, small business areas were logged. Some old tenures didn't regenerate naturally, properly, and we then put out silviculture surveys – people,

Planting near Campbell River, 1949.
NA-09733.

individuals. Or they'd have a company name, would be contracted, and they'd run lines and put plots in and see the number of trees there. The plantable spots would come out of that. Perhaps it's got 2000 plantable spots, so you order, say, 2500 trees. This is two years ahead before they go in the ground; you know what you're gonna do. Then I look at those surveys. Later on I order seed for those plots, and then at the same time they might need site prep – say piling, burning, scarifying or disking. So we finally get the trees coming, and they go into cold storage. In the meantime, you've taken the planting companies out there and you offer the contract. So they come on a certain day and we have a viewing. We go around all the blocks, and they view it. Then they put in their bids, and then you have your successful bidder ... contractor. Then as spring comes along, you already keep an eye on it for when the ground's ready. Then you issue an order to work. So they have to be ready, and they could be planting somewhere else. So you issue that, and they come along and they have quite a few planters, quite a few foremen, so then you just get them in the ground.

The grazing division also continued its considerable role within the BC Forest Service's programs, conducting reviews, range work and monitoring of the ranching industry. Certain special problems were also starting to come to the fore – some of these would influence many Forest Service activities, as a growing awareness was emerging of multiple uses in certain areas of the province. One area that featured prominently in the 1970s, the range area around Francois Lake, was highlighted as an area of concern as early as 1966 by the Prince George forest district's agrologist. He pointed out livestock's overuse of the range was affecting the area's use by later wintering deer and also affecting important shrub species. In addition, he was greatly concerned about the Peace River country, where a series of pastures had been established which required a different type of care than the open ranges.

The End of an Era

When Ray Williston ended his term as Minster of Department of Lands, Forests and Water Resources and was replaced by Bob Williams on September 15, 1972, it was the end of an era. Although policies implemented first by Orchard and then modified and added to by Williston continued for the next several years, new approaches were on the way – by Williams and later by Tom Waterland, who succeeded Williams as the minster of a newly formed Ministry of Forests. Williston had maintained the prime objectives established by the first and second Sloan Commissions and by Orchard: first, economic growth

Robert Williams, minster of the Department of Lands, Forests and Water Resources, 1973. ZZ-95146.

and support for industrial harvesting of the forest, and second, maintenance and improvement, when feasible, of a commercial forest for future harvesting. Although these remained the major objectives throughout Williston's term of office, a variety of subsidiary objectives increased in importance, including range management and improvement of recreational access to the forest. The Williston era was an era of prosperity for British Columbia, and much of that prosperity depended on a growing forest industry.

Chapter 6
Adjusting the Program

In 1972 a New Democratic Party government was elected, and the premier, Dave Barrett, handed the Forest portfolio to Bob Williams. There was immediate and negative reaction from industry to government statements that promised sweeping changes. But a review of the NDP government years suggests their new approach did not seriously disrupt the existing activities of either the BC Forest Service or the forest industry; rather, it was the fear of such disruption that created a series of problems. Certainly, there was considerable concern over the way Williams operated and considerable disagreement with many of his decisions, but nearly every issue that was of concern to industry, except the *Coast Logging Guidelines*, had begun while Williston was minister.

The coast logging guidelines were a Forest Service initiative aimed at meeting some of the new demands for environmental protection. These guidelines were intended to improve the balance of use in the forest and protect

Faller at South Bentinck Arm, 1976. I-05484.

values other than the cheapest harvest.[1] The issues of expanding parks and having more stringent environmental regulations had already been discussed, and confrontations with First Nations had been rapidly snowballing from the early 1960s on. It was in the early 1970s that all of these issues began to boil over. Environmental and other citizen action groups presented their case for the preservation of old-growth forest. First Nations argued for traditional use and jurisdiction over cutting programs in their traditional territories, and other government agencies started to exercise their muscles against the Forest Service policies that promoted harvesting based on sustained-yield formulas regardless of other concerns. Williams made several controversial moves, including implementing the coast logging guidelines and creating the Purcell Wilderness Conservancy. But in the face of severe resistance to change, he fell back on a tried and true method of shifting government policy: the appointment of a Royal Commission.

Buildings, Roads and Bridges

The Forest Service, with special funding in place (the Forest Development Fund, established in 1952), began a major program as early as 1947 of infrastructure development throughout the province. The old, inadequate Forest Service buildings were replaced, and new vehicles and equipment were acquired. The Forest Service, by locating, surveying and building roads, helped the smaller loggers and opened new forest areas.

New Office Buildings Completed 1947–1951

1947	1948	1949	1950	1951
Arrowhead Birch Island Enderby Kaslo Merritt Nakusp Oliver Parksville Princeton Vanderhoof	Numerous projects started	Aleza Lake Barriere Canal Flats Elko Edgewood Houston Mission City McBride Sicamous Sechelt Terrace Westbank	Arrow Park Columbia Hardy Bay 100 Mile Powell River Pender Harbour Port Moody Queen Charlotte City Yahk	Chase Cowichan Lake Harrison Lake Penticton

New ranger station at Sicamous, 1949. NA-09805.

The diversity of work being undertaken by 1952 resulted in a new division being formed: engineering. The engineering services division drew a series of support functions into a single program that included locating and building access roads, construction of buildings, creating a radio network, and maintaining and purchasing equipment. The new division immediately increased the hectic pace already underway in the Forest Service. In 1952 initial construction was made on seven miles (11 km) of road through the Aleza Lake Experimental Station. In the Niskonlith Forest 12.1 miles (19 km) were completed and another five miles (8 km) of right away were cleared on the Naver access road, plus the division commenced locating for a further 83 miles (134 km) of road. As well, 392 new items of equipment were purchased. The total vehicle fleet was inspected and warranted maintenance undertaken. The new purchases ranged from one new gas excavator to smaller purchases, such as 68 outboard motors.

Shovel loading a 10-ton Leyland truck, 1952. NA-13574.

Part of the building construction program had been completed prior to formation of the division, but this activity still engaged a considerable portion of the branch's resources. At least 28 different buildings were constructed in 1952, and new trailers for road crews and planting crews were built. Ten bunk trailers were built along with other trailers, for a total of 24 new trailers being put into service. Improved radio communication was increasingly emphasized, and by the end of the year an additional 109 radio units had been put into service, bringing the total to 687 units of varying sizes and types.

Seven years later, in 1959, road construction consumed much of the division's resources. In 1959, 604 miles (974 km) of road were reviewed for possible construction, a further 237 miles (381 km) had been located, and the Forest Service constructed 68 miles (109 km) of finished road. Between 1950 and 1959 over 282 miles (454 km) of road had been built. Road work in one form or another was carried out in 28 different Public Working Circles or Public Sustained Yield Units. Some projects posed special challenges. For example, there was no connecting road when the Kingcome Access road was built, and all equipment had to be transported by water. The Forest Service's marine station had converted surplus military landing barges in 1947 for just such occasions, and these were pressed into service, although the heaviest equipment was moved by commercial towing companies. The first task was to create a camp for 36 men before road construction could start. Next, the right of way was cleared up the valley. Road construction was set in motion with a Kohring shovel and Caterpillar track loaders, keeping 10 gravel trucks busy while a P&H shovel built grade, Cats levelled the ground and a grader made the final adjustments. One of the larger projects was the construction of a 240-foot (73 metre) bridge across the Kingcome River – the longest built by the Forest Service up to this time. The contractor used three 80-foot (24 metre) spans of Douglas-fir glulam.[2] The whole project took less than a year. Work started in January 1959, and 11 miles (18 km) of road were completed by November.[3]

Surveying the centre line of the Swift River access road, 1958. NA-17314.

Hayes truck at the Forest Service Marine Station, 1955. NA-15504.

Other work could not be neglected. In 1959 the Forest Service had 804 vehicles in service, ranging from small passenger vehicles to trucks with a gross vehicle weight of 40,000 pounds (18,144 kg). Although several new buildings and a number of porta–type residences were completed, the largest building project in 1959 was the construction of a seed extraction plant for the Duncan nursery. Construction was also started on an adjacent freezer facility for the extracted seeds. Radios continued to be of major importance, with 56 new AM radios being purchased along with 82 FM radios.

Difficult Times

The NDP came to power with great expectations but little in the way of a planned agenda. The Forest Service was already tentatively moving in the general direction of the NDP's pre-election promises of greater regulation of coastal harvesting, but these steps were deemed inadequate by nearly everyone except the forest industry, who began denouncing these new constraints. The coast logging guidelines were released shortly after the NDP assumed office and immediately became the principal skirmish that Williams and the BC Forest Service faced on the coast. How to deal with the coast logging guidelines and how to enforce them, what kind of leave strips (section of forest left between each cut block) to create, how to keep the loggers from using the streams for gravel and as the easiest route to haul logs from one side to the other, how to ensure that landslides and other problems associated with steep-slope logging did not impact salmon streams – all these questions assailed Forest Service staff. The downward economic turn of the early 1970s, a trend that continued into and beyond the 1980s, resulted in considerable pressure from industry to relax the new regulations. Williams and the Forest Service were caught between the rhetoric surrounding the *Coast Logging Guidelines*, their desire for improved logging techniques, and the economic reality of the 1970s, when the forest industry was still the largest provider of government revenue. Although the coast logging guidelines remained in place, great difficulties in enforcement and even interpretation (i.e., what the guidelines meant) continued throughout the period Williams was minister. There was some

A large clearcut in 1973. The Coast Logging Guidelines were an initial attempt by the BC Forest Service to come to grips with the concept of multiple-use forests. NA-29196.

The Nazko, 1977. I-05125.

support. For instance, the Vancouver forest district had promoted the coast logging guidelines, but as usual, the Forest Service remained caught between their mandate and the ever-growing collection of interested parties with other agendas, as they struggled to refine the regulations into a workable solution for everyone.[4]

Although the coast logging guidelines presented challenges for Williams and the BC Forest Service in the early 1970s, they were not the only source of difficulty. First Nations were becoming more and more forceful in their assertion of their traditional rights. Many of the problems the *Coast Logging Guidelines* were intended to fix were overshadowed by the emerging logging controversies that surrounded and continue to surround Haida Gwaii, controversies that emerged in other areas of the province after 1974. These problems were not new, and some members of the Forest Service had long argued for First Nations' rights, even suggesting that First Nations be provided with greater access to the forest resource on their traditional lands.[5] By the early 1970s First Nations' voices were becoming more insistent and reaching the most distant places, like the Nazko, for example. In June of 1973 Minister Williams received a petition to stop all logging in the

Blackwater area. This petition was, initially, from an ad hoc committee, but shortly thereafter, and more importantly, a second petition came from the Nazko Indian Band. While logging had not actually begun in the area in 1973, the Forest Service had been allocating cutting rights. The Forest Service quickly agreed to some of the petitioners' requests, such as not putting the logging road through the Nazko village, but the crisis mushroomed into a demand for a moratorium on all logging in the area and was set in the context of First Nations' traditional rights. Similar situations were occurring throughout the province. It was clear that in many areas of provincial forest the Forest Service could no longer issue cutting rights without prior consultation with local residents or else they would face regional protests, be it from First Nations, ranchers, trappers or even seasonal hunters. When the Nazko band felt they were being ignored, Chief Catherine Patrick wrote directly to Premier Dave Barrett:

> We do oppose the way development is occurring. We are prepared to agitate for our rights and our future. We request that you now protect these rights by placing a moratorium upon this area and the rest of the upper Blackwater drainage area until we can present you with a plan that will allow development without destroying our people.[6]

Sawmills at Quesnel, 1970. NA-25571.

Ecological Reserve No. 27, Yellow Pine. 1972. NA-27220.

Eventually, these concerns resulted in the Department of Indian Affairs and Northern Development also requesting a moratorium on logging in the Nazko area until a study could be undertaken. Williams agreed to the moratorium, which immediately placed the Forest Service into the thick of things. Quesnel and the sawmill industry at Quesnel, which had been requesting cutting rights in the area, declared there would be timber shortages and layoffs unless the timber was released. In early 1974 Williams approved initial logging in the area, but under strict guidelines that included special restrictions relating to water quality and stream damage, and no logging camps were to be built in the contentious area. But by May the decision had been temporarily reversed, and a further three-month moratorium was imposed on road building and logging. Following the election of December 1975 and the change of government back to Social Credit, under Bill Bennett, logging was authorized to take place in the Blackwater area in 1976.

The Forest Service's attention was now as frequently directed to non-timber issues as it was to actual timber management. In the mid 1960s, a planning file system had been implemented to deal with the non-timber issues, which involved everything from a proposed ecological reserve to new or expanded parks, recreational and other land uses or even an application for a gravel pit. Recreational use requests were especially burdensome in forests close to main highways and in areas already well utilized for recreation, such as the Okanagan. Only rough roads connected many of the lakes in the Okanagan, including Beaver Lake and the Dee Lake chain, but this did not discourage the anglers. The Fish and Wildlife Branch suggested upgrading the road, which was of a considerable concern to the Forest Service given the heavy use already made of it, but an upgrade could bring other benefits, including improved wildfire protection access. Regardless, it was clear that recreation was going to increase significantly in these areas, and the Forest Service began establishing additional recreation sites and placing map reserves on several of the more heavily used lakes.

Significant withdrawals were also being asked of other provincial forest areas. In 1974 Williams established the Purcell Wilderness Conservancy, which shook industry to its roots. Crestbrook's chief forester, the largest employer in the East Kootenays, declared that this "was a tragic event.... It's going to have a disastrous impact on us."[7] The Purcell announcement was followed in 1975 by a hefty expansion of the boundaries of Cathedral Park. In this instance, Northwood Forest Industries felt betrayed, as they had worked closely with the local parks society on the initial founding of the park in 1968. From the loggers' point of view, conservationists have found "out where all the good timber is, and said 'now we'll make a park.'"[8] The preservationists, in turn, argued that they were preserving the last wilderness areas from the roughshod logging that devastated the forest. Caught in between was the Forest Service. The legislation was relatively specific as to the limits of Forest Service authority and the activities they could engage in. And whether the Forest Service was in agreement or not, the policy of necessity was to abide by the legislation and regulations in place, until the government revised them.

As if the concerns over varied land use and water quality were not enough, the signs of a future catastrophe had already appeared in the early 1970s: that of beetle infestations. It was of particular concern east of Kelowna, where nearly 9000 acres had been infested. At this stage, the problem was primarily Spruce Beetle, but the presence of Mountain Pine Beetle was also growing. The Forest Service still thought it might be possible to control the outbreaks by clear-cutting the timber. But an added concern was that the infested areas were highly visible, and for this unexpected harvesting program, no suitable seedlings were immediately available for reforestation. The situation was severe enough that a federal entomologist deemed the problem endemic.

The BC Forest Service sought solutions to these many issues and sought to address each concern timber area by timber area. The Service's main direction was overall resource planning. The *Ministry of Forests Act* and the *Forest Act* were amended in 1978 to include formal references to resources in addition to timber, including recreation, water, wildlife and aesthetics. By 1974 John Stokes, the deputy minister, was prepared to state categorically that the Forest Service "seeks input from many agencies pertaining to forest harvesting plans." All cutting permits were being referred to Fish and Wildlife Branch, Water Rights Branch, Environment Canada and, when needed, irrigation districts, city councils or other concerned parties. District by district the Forest Service sought to develop resource folios that would identify areas of concern. Many of the difficulties arose not because of a lack of good intention; rather, available resource planning had not yet been completed in the area in question and there was inadequate understanding of the strong new demands being made on the forest by special interest groups. Mistakes were bound to occur as the Forest Service sought to balance its mandate of industrial support and long-term forest renewal with the demands of other agencies and stakeholders.

Public meeting at Prince Rupert, 1981. NA-37108.

Another Royal Commission

On June 12, 1975, by order-in-council, Dr Peter Pearse was appointed Royal Commissioner to investigate and report on the "management, regulation and use of the Forest Resources of British Columbia."[9] Approximately 15 months later, a two-volume report to the government was published. The NDP was defeated in December 1975, replaced by Bennett's Social Credit government, and Tom Waterland was put in charge of the Ministry of Forests. The government proceeded with a full assessment of the Royal Commission report and implemented the recommendations of the review committee, composed of Forest Service personnel.

In the end, each recommendation was extracted and reviewed. The committee also conducted interviews with a wide range of interested parties, but these interviews still commonly focused on what industry had to say. Although presentations to the Royal Commission were made by a variety of groups interested in issues other than logs, such as the Sierra Club and the Comox Mountaineering Club, the hearings were dominated by industry's voice. Pearse realized that most British Columbians had not been heard from. He knew and noted in his report that non-industry voices were growing more strident, but he focused his report on issues surrounding the economics of industrial use. A failure to sufficiently emphasize growing non-industry concerns to government and government's failure to grasp this critical situation resulted in cancellation of the coast logging guidelines and possibly led to the so-called "War in the Woods" of the 1980s and '90s.[10]

The Pearse Royal Commission provided baseline data on how well the previous methods of forest administration had worked. Pearse strongly argued that some prior methods – particularly that of often establishing policy by regulation rather than legislation, the method preferred by Williston – were at times, perhaps, even illegal. He also noted that broad resource planning was urgently needed and proposed an approach, for at least the short term. But Pearse remained committed to the Forest Service's view that while all values should be analyzed, the economic situation demanded the most attention. He did not, for example, support preservation for preservation's sake.[11] He did recognize that public participation in resource planning had to broaden, but he remained fixed on concepts embedded in a legal approach and recommended that local planning take place through regional district boards. This soon proved an unsuitable forum.

A New World

Proclaimed on January 1, 1979, the new *Forest and Range Act* continued many past practices but under the jurisdiction of a significantly reorganized Forest Service. This reorganization, combined with the provincial government's restraint program that began in 1983, resulted in considerable turmoil among Forest Service staff in the 1980s. Perhaps the greatest shock was the disappearance of a venerable title and job, that of ranger. The forest and ranger districts were now divided into regions and forest districts, each with a manager. The district manager still reported to the forest region, which still reported to Victoria, but much of the authority that once resided with the district forester was now transferred to the district manager.

Bob Petchie

Under the ranger district system, the assistant ranger would go out, and he might collect some fees from a scaler on a small private operation and at the same time be looking at a logging operation and at the same time be checking a culvert to see if it was clear.

Under the new structure there would be a specialist that dealt with range, there would be a specialist that would go and see if the culverts were checked, there would be somebody else to go out to see if the timber was hauled.

The objective of the reorganization was partly to move some authority away from Victoria to the regions, and even more so into the forest districts. The number of districts was also considerably reduced, from 99 ranger districts to about 46 forest districts, although the exact number varied slightly over the next several years. In some respects the Forest Service structure seemed similar to what it was previously, with merely the names changing and boundaries shuffling to suit a new era of easier access to remote areas. Still, this was not an easy time for staff. Change was not new, though. The 1974 certification of the British Columbia Government Employees Union (BCGEU) had changed many relationships, as had a number of past Forest Service reorganizations, including the large-scale ones of 1952 and 1958. More recent changes included the formation of a Cariboo forest district in 1972. The changes of 1979 and 1980, however, were seen by many as traumatic and resulted in numerous early

Cariboo District office employees, 1973. NA-30057.

retirements. Many staff continued to regret the loss of positions like district forester (now replaced by regional manager) and ranger (now replaced by district manager), and they disliked the transformation of the Forest Service such change represented.

The new organization rearranged reporting structures as well. Headquarters was now divided into five divisions, and each division had five branches – except for Operations, to which the six regions reported. In addition, two separate branches reported directly to the deputy minister: Information Services and Strategic Planning. This organizational structure remained intact until 1984–85, when Victoria headquarters was organized into three divisions and the branches were reduced from 17 to 15. In addition, Range now became a branch reporting directly to the deputy minister. In 1986–87 headquarters was again reorganized, when the responsibility for Crown lands was transferred to the Forests Ministry, which became the Ministry of Forests and Lands.

One major shift was the change from Public Sustained Yield Units to Timber Supply Areas. This change started in 1978, with the intent of redefining all the Public Sustained Yield Units as Timber Supply Areas by 1980. The new boundaries were designed to encompass areas seen as economic watersheds. The sawmills of Quesnel, for example, would all be supplied from the same Timber Supply Area rather than from several Public Sustained Yield Units. A Timber Supply Area (TSA) was defined as "an area of Crown land designated by the minister of forests in accordance with the *Forest Act* and managed for a range of objectives including timber production."[12] A recent definition has described TSAs as follows:

> A Timber Supply Area is a designated area established by the Ministry in order to practice sound, integrated, resource management principles to improve the allowable annual cuts. TSAs were originally defined by an established pattern of wood flow from management units to the primary timber-using industries. They are the primary unit for allowable annual cut (AAC) determination.[13]

Although part of the objective was to simplify access to logs, greater care was now also to be taken to ensure that other values such as recreation were being analyzed and considered. In addition, regular assessment and review was to be undertaken for each TSA.

Sawmills near Quesnel, 1970. NA-25574.

Another major change was that new authority was gradually shifted to the revised districts from the region. Direction of the silviculture program was now under the control of the district rather than the region, and much of the work for timber administration was also transferred to the new districts. The district managers found themselves with a variety of new roles and responsibilities, as well as the difficult task of consolidating a wide range of personnel, who had previously operated relatively independently, into new working relationships. The new Port Alberni forest district, for example, encompassed the old Parksville, Alberni and Tofino ranger districts, and it was several years before the various locations were consolidated into one central location in Port Alberni. In 1997 more changes brought about the creation of the South Island forest district, which combined the Port Alberni and Duncan forest districts. The Duncan forest district had been a union of the Duncan, Langford and Lake

The effects of logging on fish habitat, 1980. The BC Forest Service, Canadian Department of Fisheries and Oceans, and MacMillan Bloedel began an intensive study of logging and fishery integration at Carnarvon Creek, Vancouver Island, in 1970. These studies contributed significantly to the *British Columbia Coastal Fisheries-Forestry Guidelines* implemented in 1987. NA-35998.

Cowichan ranger districts. Similar amalgamations were being made in other districts around the province throughout the 1980s and '90s.

During the period from 1979 to about 1987 much effort was focused on managing the Forest Service's new responsibilities and organizational structure. One of the larger programs that was gradually implemented was the Small Business Forest Enterprise Program, whereby selected areas of each region were made available for small operators to purchase timber. A major concern raised by the Royal Commission had been the concentration of more and more forest under the sway of the larger integrated forest companies. Pearse believed that a more diverse forest industry would mean a stronger industry, and so the committee analyzing the Pearse report proposed that something be initiated to give small loggers a better chance to secure timber. After some study, and with the support of Minister Waterland and Deputy Minister Mike Apsey, the Small Business Forest Enterprise Program (now called BC Timber Sales) was implemented in 1980 on an experimental basis. As with any new program, a few unexpected situations arose, but modifications were rapidly made to improve the operation.[14]

Other activities also came under scrutiny as the Forest Service settled into its new routine. For instance, a report on revenue administration in 1985 noted that although some control over billing and accounts receivable had already been downloaded to the districts, even more responsibility should be transferred.[15]

Additional players external to the Forest Service were having an impact as well. In 1985 the government appointed a Wilderness Advisory Committee to study areas that should be retained as wilderness, and in 1987 the *Forest Act* was amended to recognize this form of preservation. Emerging concepts were shifting Forest Service priorities. These sometimes came with new resources, but usually they did not. One of the many new tasks the districts undertook was to

> identify areas having wilderness values and rank these areas (and areas identified by other agencies, licensees or the public) for further review using the wilderness ranking guidelines.... participate in assessment and designation processes for identified wilderness study areas, and prepare Wilderness Management Plans for designated wilderness areas and monitor approved Wilderness Management Plans....[16]

During much of the early 1980s the Forest Service was focused on the aforementioned internal reorganization but also on the growing softwood lumber export disputes. Considerable attention at the senior levels was devoted to establishing equilibrium in both of these activities. This period brought at least two additional major concerns as well: integrated resource management and reforestation policies and funding. Other concerns also occasionally shifted attention toward the growing concerns of special interest groups, such as at Riley Creek on the Queen Charlotte Islands (Haida Gwaii), where a dispute between the federal Department of Fisheries and Oceans and the Forest Service over logging and the destruction of fish habitat finally had to be resolved at the ministerial level.

For the BC Forest Service, however, the decade ended on a better note. Following a meeting of the Canadian Council of Resource Ministers in 1980, the Canadian Forest Service developed a Forest Strategy for Canada, and from this emerged two Forest Resource Development Agreements that provided substantial new funding for each province, particularly for British Columbia. Funds became available in 1986 and were intended to help resolve the most pressing issue identified in the 1981 Forest Strategy: the restocking of harvested lands that, for a variety of reasons, were not regrowing merchantable timber as expected and were termed NSR (not satisfactorily restocked). The Canadian Forest Service considered this restocking to be essential if Canada was to maintain a healthy and sustainable forest industry.

Peter Poland: On Supervising Tree Planting in the Clearwater Area
Tree handling became a big thing, making sure. They found that, you know, if you planted trees and didn't handle them correctly – so if the trees were getting too hot during the day and they weren't tarped off properly, they still looked fine – you know, you'd find out the next year they were all dead. I wonder why. So tree handling was a huge issue in making sure that the tree planters were putting the caches in creek beds or along creeks and nice cool places and shaded and tarped. You know, that was part of our job, to make sure that they weren't sitting in the back of a truck, you know, in the middle of the afternoon getting up to, you know, 40 degrees.

The first Forest Resource Development Agreement, signed in 1985 and underway in 1986, was to last five years and was followed by a second agreement extending the life of the program a further five years. In addition, the province amended the *Forest Act* in 1987, making reforestation a legal obligation of those who harvested timber. Major licence holders now had to do this work at their own expense, while the ministry undertook the work in the Small Business Forest Enterprise Program (although the logger was billed for the costs of the work). Initial federal funding for the backlog work was $300 million, and the Forest Service immediately moved into high gear, dramatically increasing tree plantings and other silvicultural work.

Peter Poland, Working as a Silviculture Officer in Alberni

The other part of my role was to look at silviculture prescriptions, which in those days … were part of the legislation. And those were basically prescriptions done by professional foresters to say "This is what would happen after" or how cut blocks would be logged and how they would be reforested and how the brush would be managed and all that kind of thing. So I would review those and see whether or not they – the stocking – was a fit with the biogeoclimatic zone. So it was a bit of a technical thing. We'd often go out before the blocks were logged and walk with the licensees and check certain things and come up with some ideas. And so it – in the 1991 to '95 era things were moving – was starting to pick up and there was lots of activity. So it was quite a busy job, and I enjoyed getting out and seeing the district, seeing…. We saw lots of things that might, you know, be problems in the future in terms of, you know, such as how steep the slopes were and, making sure we protect the streams and gullies. So there was lots of that, and working with the research people.

The BC Forest Service estimated that by the end of 1988, the not-satisfactorily-restocked timber lands had been reduced by nearly 25 per cent and a vast amount of additional work had been undertaken that provided considerable employment to tree planters, forestry crews and many others. Work under this program continued on into the mid 1990s.

Creating the New Forest

Forest Branch staff by at least the early1920s were concerned about how the forest was going to be replaced once harvested. The Fulton Royal Commission of 1910 had pointed out that silviculture should be part of the branch's program. The successful silvicultural management of a virgin forest should result in a complete and even restocking of cut-over areas with young growth of the best species that the area is capable of producing. In practice, this result is sought by rules planned to secure a suitable seed bed for the desired species and an abundant supply of the right kind of seed.[17]

Several early Forest Branch foresters expressed their frustration at the industrial practice of harvesting without regard to how the forest would regenerate, and the branch began documenting wasteful logging methods. Growth and yield studies began in a small way in 1920 and 1921, as the branch began to investigate regeneration on cut-over lands with the aim of eventually establishing a silvicultural policy. In 1922 a series of plots were demarcated and a silvicultural description prepared for each plot. Progress was slow but steady. In 1923 Chief Forester Caverhill noted that

> No matter how extensive they may be, the exhaustion of our virgin stands is only a question of time, due to increasing demands of the lumber market. Depletion in the East and South and constant shifting of the centre of lumber production to the Pacific coast has focused attention as never before on the forests of British Columbia. Forests are a crop and, unlike minerals, are renewable, provided that the natural laws governing their reproduction and growth are understood and followed.[18]

The establishment of a small experimental nursery on Shelbourne Street near McRae Avenue in Victoria in 1926 heralded a greater effort directed toward reforestation. Initially 30 beds of Douglas-fir and Sitka Spruce were planted. The nursery continued to increase in size until it was closed in 1932 and the land sold. In the meantime, a nursery at Green Timbers in Surrey was established in 1930 and became the Forest Branch's main nursery for the next several decades. Demand for seedlings for replanting grew significantly after the implementation of sustained yield, and nurseries were established at a number of other locations.

Nurseries

Name	Nearest Community	1967 seedlings in millions	Year opened	Year closed or sold
Shelbourne St	Victoria		1926	1932 or 1933
Green Timbers	Surrey	7	1930	1998 or 1999
Campbell River	Campbell River	12	1939	1988
Duncan	Duncan		1943	
Elko	Elko		1949	
Wycliffe	Cranbrook		1950	1967
Koksilah	Duncan	8.6	1963	1989
Chilliwack	Chilliwack	12	1967	1988

Red Rock	Prince George	15	1967	1988
Surrey	Surrey		> 1968	2004
Skimikin	Salmon Arm		1972 or 1973	2002
Harrop	Nelson		1974	1988
Telkwa	Telkwa	1		1988
Thornhill	Terrace		1985	1988
Vernon	Vernon			1988
Brennan Lake	Nanaimo?			
Rayleigh	North Thompson	2.5		

Cone collection was also an important activity, and the cones collected were sent to the Duncan Seed Centre for extraction and processing. Seeds were collected in various ways, including shooting cones off the tree (although this technique was only used to collect for

Truckload of cones at the Duncan nursery, 1975.
NA-31452.

research purposes) and picking cones from helicopters. The seed crop was monitored closely but was dependent on local conditions. Cone production was also cyclic, and really good crops sometimes only occurred every 14 years. In an effort to ensure adequate and high quality seeds, the Forest Service established several seed orchards, beginning in 1963. As a report on the Kalamalka seed orchard noted in 1992: "Seed orchards are a key element in a successful tree improvement program. Their purpose is to mass produce high quality, genetically improved and biologically diverse seed for reforestation requirements."[19]

Reforestation was much more than simply growing seeds into seedlings, lifting them and then replanting in harvested or burned areas. Silviculture treatments varied substantially, although some methods were used more substantially than others in a given period. When sustained yield was initially introduced, silviculture was relatively simple in approach – basically, what method should be used to cut the forest and, once it was cut, how was it going to be regrown for future harvesting? This was most commonly achieved by leaving seed trees and, for a time, one method used extensively was scarification, which prepared the landscape and soil for the new seedlings. In addition, brushing and weeding, spacing, pruning, thinning and even spreading fertilizer might be required. Experiments were conducted to protect

Thinning a Lodgepole Pine stand, 1983.
NA-39047.

the seeds and seedlings and ensure a new forest would grow. Protecting seedlings and seeds was important. As a study published by Forest Service genetics scientist Alan Orr-Ewing in 1950 noted, one of the major problems in the natural restocking of cut-over areas was the Deer Mouse. Orr-Ewing described these mice as "avid seed eaters" that happily moved into harvested areas, thereby effectively checking natural reproduction.[20]

New techniques were also developed and studied. Scarifiers from Sweden were tested, celery planters were transformed into tree seedling planters, nurseries and new techniques for transplanting seedlings also were tried. For example, Jack Walters of the University of British Columbia developed a transplant method using plastic tubes or bullets. Beginning in the 1960s and 1970s, seedlings were being planted by the mud pack method, wherein each seedling is pressed into a cigar-shaped capsule of mud and clay. A hole can then be made with a planting tool called a dibble and a seedling's roots placed in the hole. In the early 1970s the BC Forest Service and the Canadian Forest Service began experimenting with seedlings grown in styroblock containers. The first experimental planting occurred near Houston in 1970. Success meant increased dependence on the styroblock system, initially at the Duncan nursery and then also at Surrey's Green Timbers nursery.

Greenhouses at the Duncan Nursery, 1976. NA-32088.

Harvesting grew by leaps and bounds, and while there was still some belief that natural regeneration might be able to provide much of the needed restocking, a growing awareness developed that this was not going to be the case. Reforestation issues concerned the Forest Service for decades, but with the new post-World War II mandate it became a critical activity.

Replanting was only one part of a greater program aimed at regenerating the forest. District silviculturists worked to ensure that harvesters left seed trees standing and that the ground was prepared for regeneration. They monitored areas where natural regeneration was not working and then arranged for the planting of seedlings. Planting trials had started in the 1920s, and as the need for seedlings continued, the Forest Branch expanded its nursery program from the nursery at Green Timbers to include a nursery near Campbell River that went into production in 1940. A nursery near Duncan was also developed. The Forest Branch had increased planting in selected areas during World War II by using workers from the federal alternate service workers program. One issue, aside from securing sufficient workers, was in the older logged areas, which were often not suitable for replanting due to the brush and deciduous growth that had sprung up. Much of BC's coast was in this condition. The importance of ensuring a future crop was recognized, and in late 1946 the reforestation

Planting seedlings at Jordan River, 1975. NA-31481.

division separated from the forest economics division. The new division was initially concerned primarily with nurseries, seed collection and planting. Silviculture, however, was being investigated by the research division and was implemented and administered by the forest districts as the various managed forest areas were created.

In 1980 the Nelson forest district prepared a five-year plan as part of the five-year *Forest and Range Resource Analysis*."[21] In it they identified their silviculture objectives. Silviculture was still preoccupied with producing timber for future harvesting, or, as the report noted, "the basic program is concerned with the maintenance of the natural productivity of the accessible forest sites."[22] The program was divided into two parts: basic and intensive. The basic program focused on cone collection, regeneration surveys, site preparation and survival studies. The intensive program focused on improving the allowable annual cut, which included review and planting of backlog areas, as well as stand tending such as juvenile spacing. The ministry engaged in all silviculture work on timber licences. By the late 1980s much was changing. A larger portion of the work and expense was being shifted to the licensees by 1987, and by the 1990s the perception of what silviculture encompassed had significantly broadened.

Lifting seedlings at the Duncan Nursery, 1955. NA-15170.

Henry Benskin, as Silviculture Branch director, was well aware of the new forces coming into play and worked toward their inclusion in Forest Service planning. As he told a silviculture conference in Toronto and subsequently wrote in 1993:

> In the past, the basic question in forest planning was: How do we get the logs out without harming the other resources? Today, the question is evolving into something more like: How can we get the logs out in concert with good management of the other resources or resource values? In part, this is happening now, but the complexity of putting it into practice is enormous in the context of managing with an appropriate emphasis for ten or even 15 values all on the same piece of land. In the future, silviculture will become an ever larger player in achieving such a result.[23]

The private sector also began to undertake more of the reforestation program after a change in legislation in 1987. Prior to 1987 nearly all replanting was funded by the ministry, but by the mid 1990s only a tiny portion was ministry-funded.

In order to meet the objectives of silviculture, especially those as outlined by Henry Benskin, a number of new silvicultural approaches have been tried, such as shelterwood or seed tree selection. In the last decade, however, another approach has become the main means of achieving the broader objectives desired in today's forest: variable retention. Emerging as an alternative to traditional silvicultural practices in the 1990s, and especially promoted by J.T. Arnott of the Canadian Forest Service and W.J. Beese of MacMillan Bloedel, this practice is based on the idea of retaining significantly more forest after harvesting and thereby retaining sufficient diversity so that effective forest regeneration occurs. The concept of variable retention has now come into widespread acceptance, although it is still most often only used in coastal BC. As described in a ministry publication, the objective is that

> Retention areas should be designed to retain the natural range of stand and forest structures, maintain natural eco-system functioning and biodiversity, provide habitat connectivity over the landscape, and supply natural refuges for the survival and dispersal of species after harvesting (e.g., wildlife tree reserves).[24]

The 1970s and 1980s were a time of considerable adjustment for the BC Forest Service. New demands, experiments in regulation and a greatly changed organization all created major upset. Many disheartened staff left. Rangers – who had long considered themselves the elite of the Forest Service, even when working under the heavy hand of the district forester and the even heavier hand of Victoria – now had to apply for new jobs. While some rangers became district managers and gained even more independence than they had previously enjoyed, many felt the new jobs (such as operations managers) reduced them in status. Those assistant rangers longing for a ranger's job now found that option no longer existed. The possibility of sweeping policy changes that emerged in the early 1970s disappeared in the 1980s, and except for the major shifts toward privatization and moving costs onto the licensees, programs by and large continued on. Although the 1970s and '80s were a difficult 20 years, the 1990s and the subsequent decade would abruptly bring even more changes that shocked and reshaped the Forest Service and its role in British Columbia.

Ranger station at Mission, 1950. Each ranger operated quite independently in his own little kingdom, with only occasional interference from the district forester or head office in Victoria. NA-10085.

Chapter 7
A Remodeled Forest Service

Since 1912 the BC Forest Branch and Service has coped with numerous changes in organization, policy, industry and government. During the last 20 years, leading up to 2012, the pace of change increased dramatically. New programs, new threats and regular reorganizations have greatly affected the way the BC Forest Service manages its mandate. Beginning in the 1940s the Forest Service entered a period of growth, and significant staff increases helped the organization to meet its enlarged mandate. But the ministry has since undergone a regular series of staff reductions, with major reductions occurring in 1983, 1987, 1994, 1998, 2001 and continuing to the present day – all while workload has continued to increase. The organization's mandate has, in many respects,

Forest ranger staff at Golden, 1948. NA-09362.

continued unchanged since 1912, but considerable refinement has gone into how it is worded, interpreted and implemented. Overall, the Forest Service has remained focused on managing Crown-owned forests in the best interests of the people of British Columbia.

In 1910 it was described as follows:

> Under the Minister the Chief Forester should direct the Forest Department in the management of all Crown timber lands; in the control and development of methods of protecting forests from fire and other damage; in the regulation of the cutting of timber and the prevention of waste; in the care for, and maintenance of, forest cover and the prevention of soil erosion; in the preservation of the young growth that is the future timber supply; in the replenishing of depleted forest regions; in the maintenance of timber areas permanently dedicated for park or other purposes; and in all other work of the Department.[1]

And in 2013:

> The purposes and functions of the ministry are, under the direction of the minister, to do the following: (a) encourage maximum productivity of the forest and range resources in British Columbia; (b) manage, protect and conserve the forest and range resources of the government, having regard to the immediate and long term economic and social benefits they may confer on British Columbia; (c) plan the use of the forest and range resources of the government, so that the production of timber and forage, the harvesting of timber, the grazing of livestock and the realization of fisheries, wildlife, water, outdoor recreation and other natural resource values are coordinated and integrated, in consultation and cooperation with other ministries and agencies of the government and with the private sector; (d) encourage a vigorous, efficient and world competitive timber processing industry and ranching sector in British Columbia; (e) assert the financial interest of the government in its forest and range resources in a systematic and equitable manner.[2]

In 2012 the BC Forest Service merged into a superministry, the Ministry of Forests, Lands and Natural Resource Operations, and continued striving to deliver in its purpose. But it is not the same organization that was formed in

Prince George Region district managers, 1982. NA-37509.
Regional and district managers, 1980. NA-37509.

1912 or even that of 1945, when the name BC Forest Service officially came into being. Even the fundamental reorganization in 1979, following the Pearse Commission, did not have as dramatic an impact on the BC Forest Service as did the changes of the last several years.

Creating the Code

Many of the changes in how the Forest Service operated began in the decades following 1945 and accelerated after 1979, as multiple use of the forest – or, as Finn McKinnon reworked the concept, integrated forest management[3] – emerged as the crucial policy trend. Land use issues appeared with increasing regularity, tighter and tighter control over environment-related activities aimed at ensuring the least damage possible, and questions concerning First Nations' rights became of even greater concern. During the early 1980s, ministry executives were focused on the major ministry reorganization and on dealing with the industrial economic crisis caused by tariff controversies with the United States. Despite these issues, progress continued in the areas of reforestation, integrated resource planning, forest ecology, research, biogeoclimatic ecosystem classification and fire protection. By the late 1980s, as the new organization had settled into its new routines, even the tariff problem almost became routine, and this allowed for progress in other areas of increasing alarm, including reforestation. The Forest Resource Development Agreement with the federal government provided funds that were used to address the growing backlog of forests that, due to fire and other setbacks, had not satisfactorily restocked. A start was also made in deciding how to cope with the massive pine beetle catastrophe that was unfolding in the province's interior. As the 1990s advanced, great change was in the wind. Cooperative regional planning had begun as early as the 1960s, but for the next decades the Ministry of Forests expended so much effort responding to various crises (or perceived crises) that little time remained for future plans. Pearse noted this in a report in 2001:

> My investigations over recent months have left me with the impression that everyone involved in the forest industry, and especially the leaders of timber companies, unions, contractors and the Ministry of Forests, have been preoccupied over the past decade by one crisis after another. The distraction of immediate problems has made it difficult for them to plan for the future, to think about where they would like to be a decade from now, and to plan strategically how to get there.[4]

But Pearse also believed everyone was ready for change, and certainly more changes were coming.

Two of the major initiatives in the early 1990s were the *Forest Practices Code of British Columbia Act* and the Commission on Resources and Environment (CORE). Although both were introduced by the NDP government that took

The forest practices code resulted in new sets of guideline and manuals such as this one from 1999 for watershed assessment. http://www.for.gov.bc.ca/tasb/legsregs/fpc/fpcguide/wap/WAPGdbk-Web.pdf

office in 1991, these initiatives were the result of several years of agitation and discussion. In 1989 the Social Credit government had tasked the British Columbia Forest Resources Commission with a complex assignment: to study the current condition of the forest and make recommendations. The commission turned in its report in 1991. Among its urgent recommendations was that the multitude of regulations and acts related to forestry be consolidated into one code. The report, although fairly thin on exact details, did highlight some extremely important issues. For example, it noted that

> In coastal regions, harvesting techniques were historically based on technical and economic criteria, rather than on ecological, silvicultural and aesthetic criteria. Harvesting policies and philosophies have been kept separate from forest management policies and philosophies. As a result, management for other values – now considered important by society – was not actively considered until after the harvesting activities had already taken place.[5]

The report was followed by the release of a discussion paper and subsequent committee meetings that led to the preparation of a code intended to achieve many of the report's various objectives. In late 1993 two more discussion papers were released and work began on preparing the legislation that came into effect June 15, 1995. The *Forest Practices Code of British Columbia Act* replaced six federal and 20 provincial statues, 700 regulations and over 3000 guidelines. New guidebooks, regulations and the legislation all required extensive work to prepare and then needed to be explained. Forest Service staff needed to be trained in its enforcement, and loggers needed to be educated as to what the new rules would mean for their work. As Marty Osberg and Brian Murphy, both BC Forest Service employees, wrote in 1996:

> The Forest Practices Code will significantly change the way the forests in British Columbia are managed. The emphasis will be on achieving stewardship of forests based on an ethic of respect for the land. The vision of the Code is one of sustainable use of the forest we hold in trust for future generations. The Code recognizes that realizing the benefits that can be derived from forests requires active management and the constant evaluation and balancing of values to meet the economic and cultural needs of people and communities. The Forest Practices Code provides the tools, mechanisms, and legislated authority required to achieve the vision of sustainable forest use.[6]

Yet even once staff were trained, this was only the start of implementing a massive program that had some impact on nearly every Forest Service employee. Analysis of logging practices to evaluate compliance, revisions to logging plans and new silvicultural prescriptions were all required.

The Manual

Forest Service activities were governed by legislation and regulation, the details of which were in the *Forest Service Manual*. By 1931 both management and operation manuals were in place for the guidance of staff. The procedures outlined were to be rigorously enforced – unless they proved unworkable, and then Victoria modified them. Periodically, the manual was revised and reprinted, and in the intervals between printings it was modified by the circular letters. Each office was provided with a manual, and each staff member was expected to know the manual as it applied to them. The regular changes to the manual were often a result of new legislation or based on difficulties encountered in the field. Frequent back and forth discussion ensured a correct understanding of legislative changes and how a particular situation should be dealt with.

The manual was so detailed that it even instructed staff as to what form a letter should take – for example, a letter appointing a lookout man or as to what food rations should be issued or how many nails were required to install 1000 shingles. As useful as such detailed instructions might prove to be, it was the broader directives that dictated what actions rangers and district officers might take. *Instructions to Forest Officers*, for example, issued in 1951 in regard to timber sales, starts off by citing the relevant sections of the *Forest Act* and then proceeds to delineate each activity, from timber cruising (an estimation of the harvestable timber) and advertising to the final disposition of the sale. Usually, the detailed instructions were put in the manual, sometimes in the attached circular, but additional publications were occasionally necessary when the topic was broader. The *Forest Service Manual*, for example, instructed that a timber sale be cruised "in such a manner as may be directed by the Chief Forester." [7]A separate cruising manual was then used for the details on how to cruise a sale. The circulars addressed specific issues such as how to charge cruising costs and when it was appropriate to accept an applicant's cruise.

The latest resource management manual was updated in 2009, although some sections still remain to be written. The most recent management services manual was implemented in 1993 and has been continually

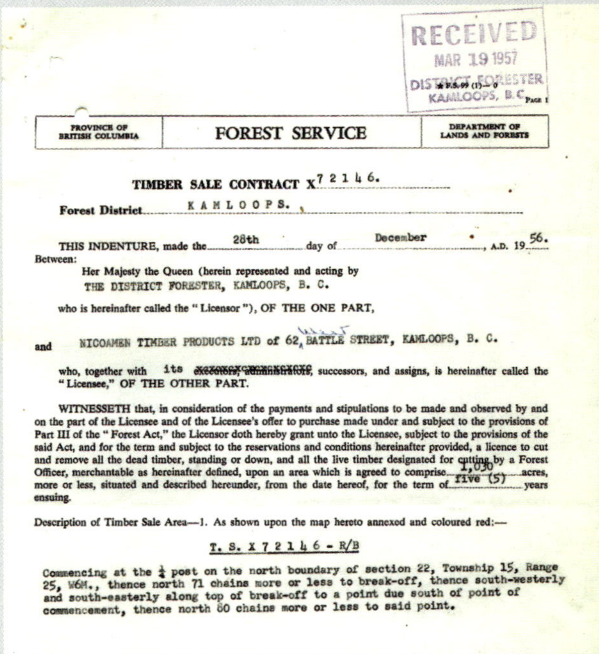
RECEIVED
MAR 19 1957
DISTRICT FORESTER
KAMLOOPS, B.C.

F.S. 99 (1)—0 PAGE 1

PROVINCE OF BRITISH COLUMBIA	FOREST SERVICE	DEPARTMENT OF LANDS AND FORESTS

TIMBER SALE CONTRACT X72146.

Forest District KAMLOOPS.

THIS INDENTURE, made the 28th day of December, A.D. 1956.

Between:

Her Majesty the Queen (herein represented and acting by THE DISTRICT FORESTER, KAMLOOPS, B. C.

who is hereinafter called the "Licensor"), OF THE ONE PART,

and NICOAMEN TIMBER PRODUCTS LTD of 62 West BATTLE STREET, KAMLOOPS, B. C.

who, together with its ~~executors, administrators~~, successors, and assigns, is hereinafter called the "Licensee," OF THE OTHER PART.

WITNESSETH that, in consideration of the payments and stipulations to be made and observed by and on the part of the Licensee and of the Licensee's offer to purchase made under and subject to the provisions of Part III of the "Forest Act," the Licensor doth hereby grant unto the Licensee, subject to the provisions of the said Act, and for the term and subject to the reservations and conditions hereinafter provided, a licence to cut and remove all the dead timber, standing or down, and all the live timber designated for cutting by a Forest Officer, merchantable as hereinafter defined, upon an area which is agreed to comprise 1,030 acres, more or less, situated and described hereunder, from the date hereof, for the term of five (5) years ensuing.

Description of Timber Sale Area—1. As shown upon the map hereto annexed and coloured red:—

T. S. X 72146 - R/B

Commencing at the ¼ post on the north boundary of section 22, Township 15, Range 25, W6M., thence north 71 chains more or less to break-off, thence south-westerly and south-easterly along top of break-off to a point due south of point of commencement, thence north 80 chains more or less to said point.

First page of Timber sale contract TX72146 in 1956.
Weldwood Collection.

amended since. A recent amendment, for example, refers to *Cruise Base Cutting Authority Policy* and *General Procedures in Mountain Pine Beetle Timber.* These are only two of a multitude of specialty guidebooks that cover everything from interior timber appraisal as compared to coastal appraisal to financial procedures. The revised Forest Practices and Investment Branch silvicultural manual is now available only online. Its purpose is as follows:

> The **purpose** of the manual is to promote best practices and to document formal procedures. Its **scope** is limited to those activities under the mandate of the Silviculture Practices Section of the Forest Practices Branch. This mandate broadly includes all activities related to forest establishment, forest health and forest stand management.[8]

The manuals, instructions and guidelines served several purposes. Initially, they were designed to tell Forest Service employees what they should be doing and how to handle any given situation. While that purpose still holds, the documents now also tell those using provincial forests what their obligations are.

Paper Trail

The Forest Service, from the formation of the Forest Branch in 1912 until today, has in many ways run on paper. Numerous forms were essential to managing British Columbia's forests. Correspondence and maps all had to be prepared, documented and filed. Forest inventories, scaling reports, lists of timber licences and who the licensees were, inspection reports and cutting allocations, as well as reports on sawmills, loggers, fire protection, fighting fires and pest management had to be submitted. Accounts books showing revenue and expenses, memos, management plans, protection plans and grazing plans were all vital to the administration of BC's forests. The Forest Service's management manual and the rulings of the forest counsel (the ministry's lawyer) resulted in a complicated but essential set of documents that guided operations. The chief forester and other staff at Victoria communicated to the district foresters, often in the form of circular letters. These could be on special topics or might be amendments to the parts of the forest manual. The district foresters often communicated in a similar fashion to rangers and other staff. As well as their regular paperwork, rangers and district foresters submitted annual reports. The chief forester then had an annual report compiled and published by the public relations and education division.

W. Henning, assistant district forester in Prince George, at his desk, 1952. NA-12529.

Bob Perchie: Retired Forest Officer
So my job as a dispatcher was to do all of the office stuff. The ranger station (about 1970) was on about two acres of land right on Stuart Lake. There was a new building plus a number of houses for ranger staff positions, because in a ranger district you moved around every two or three years. They often offered housing, particularly in the more remote locations. It took a couple of, I think it was two, months or something before a Forest Service trailer became available on the ranger station for us to move into. And so we were located right on the ranger station, along with three or four other families that lived there, including the ranger. My duties were to answer the phone, to do the typing, to wash the floors, to keep the coffee fund honest and operable and do other things that was expected of me. And Bill Brash, the ranger, he expected the floors to be washed once a week. One thing I was able to do, I was a pretty good typist. I didn't realize that, but I had taken a typing course in grade eight or grade nine or grade 10 – somewhere back then. And there was a lot of typing.

Glen Heyes: Retired Agrologist, Merritt
If there was something really controversial you had to do, in those days we'd send the carbon copies of letters and we'd – you know, we'd have a copy for our file and the original to whoever was the rancher or whoever was receiving it, and a copy to Victoria. So what we used to do, if the letter was controversial, we'd generate the letter, we'd send the Victoria copy off but hang onto ours. Wait and see what the reaction was. And if it was fine, then we'd send out the original, you know. And if hell would break loose, we'd say, oh, well, that's okay, we can fix that, we'll get the letter back or, you know.

Dianne Clearwater: Executive Assistant, Kamloops
We used to have huge big ledgers that we would work with, and you would put every cutting permit in it and how many inspections you'd done. And so the whole history of the cutting permit would be in these ledgers and who it was that had gone out on the inspection and what they found.

The initial phase under the Forest Practices Code required a company requesting harvesting rights to prepare a forest development plan. This plan needed to cover every aspect, including sensitive sites, a visual impact study, public meetings and instructions to address First Nations concerns. The plan also had to relate to broader regional plans similarly identifying all criteria that

Logging truck on a bridge, South Bentink Arm, 1976. I-05480.

might impact the Forest Development Plan, such as recreation sites. Although a broad range of legislation and regulation had been consolidated into a single act, the planning processes had become much more difficult. Previously, such plans had been incorporated into the licensing documents. But now, significant public input and much wider distribution of information was required before the Forest Service could move into working with loggers to create management and detailed cutting plans. Plan approval was at the district manager level and under the scrutiny of the newly established Forest Practices Board (created in 1995). Although the new code rigorously addressed the environmental issues that had been raised, it also meant that six operational plans had to be approved before logging could take place. A 1997 report suggested that between 1992 and 1996 the code led to a 60 per cent increase in forest industry costs.[9] As a result, some streamlining did occur following 1998 – for example, the planning process was reduced to three reports.

New Help for Growing Problems

When the NDP came to office in 1991, they faced a number of serious problems in the forest industry. The War in the Woods had erupted in the 1980s, sparked by proposals of MacMillan Bloedel to log Meares Island on the west coast of Vancouver Island[10]. The BC Forest Service was again caught in the middle on a number of major issues – a result of past government policy and the legislation the Forest Service was obligated to work under. Not only were dramatic environmental issues confronting BC's government and Forest Service, but the tariff issue with the US remained a problem, increased international competition was undermining economic recovery and current harvesting levels could not be sustained. Although considerable restocking had occurred through the two agreements with the federal government, the Mountain Pine Beetle was an ever-growing menace. A dramatic reduction of available, harvestable timber would significantly hurt the forest industry, and the pine beetle issue also contributed to increased alarm over wildfire protection.

Between 1991 and 2001 the NDP government developed a series of legislative reforms and some reorganization of the Ministry of Forests to meet the growing problems. CORE and the Forest Practices Code were only two of their approaches, but all of the changes meant that the Forest Service had to provide support and once again change their organization and activities to meet the new demands (implementation of the code being the most dramatic shift). One change was that in 1992 the ministry began a new initiative to monitor the

Pine-beetle survey.
NA-39049.

province's timber supply. Regular reviews and assessment of timber supply, in order to aid in determining the allowable annual cut, were now to be regularly scheduled. The review would include public discussion and input and provide a 250-year estimate of timber supply. In order to provide more help for resolving some of the growing forest-related issues, on April 3, 1993, Premier Mike Harcourt announced he was forming the Forest Sector Strategy Committee. In 1994 the committee released a Forest Renewal Strategy and announced the creation of Forest Renewal BC as a Crown corporation. This corporation provided the Forest Service with an opportunity to greatly expand some of its programs, especially in the area of silviculture. Financial assistance was also provided for other ministry programs, including recreation, but the central interest of Forest Renewal BC was overall forest health.[11]

The government also made several significant organizational shifts within the Forest Service. First, although Protection Branch continued to report as a program through an assistant deputy minister, the responsibility for wildfire management was removed from the regions and districts and given directly to Protection. Protection now directed their own staff and had complete management of wildfires. Additional fire bases were established, and training

Rapattack training at Little Fort, 1978.
NA-34412.

and placement of rapattack crews expanded to quickly reach and extinguish difficult-to-access fires by rappelling from helicopters. For fires where that was not possible, extra resources were called upon through a partnership program with the ministry staff.

Wildfire

"Your Commission (1910 Fulton Report) note with pleasure that in this Province the Government is strongly impressed with the need for forest protection. It began some four years ago by making an appropriation in the estimates for the purpose, and by placing men upon patrol in different forest districts."[12]

Protecting the forest from fire was in the best interest of all parties involved in forest activities, and to the government it made economic sense. Fire protection was one of the original tasks of organizations that merged into the 1912 Forest Branch, and it has remained one of the Forest Branch and Forest Service's central concerns. Other potentially destructive forces were also alarming, like insects and disease, but wildfire dominated the government's attention and effort. Fire was destructive not only to the forest. It had the potential to destroy much more, as we have seen recently – for instance, the growing communities scattered around the province.

Hicks lookout, Prince Rupert District, 1951. Left to right: C.D. Orchard, R.G. Minger, C.L. Gibson. H-02958.

Detecting and fighting big fires was a heavy investment in all resources. The objective was to locate and get to the fire before it became a massive conflagration. In 1914 the chief forester noted that

> In a forest Province such as British Columbia the prevention and suppression of fires is of universal concern and importance. Every interest, private, corporate, municipal, Provincial, and Federal, is affected in some way, and mutual assistance and co-operation is necessary if the efforts of fire-protection are to be effective. It is to say that in no other Province or State is a more active interest taken or more help given by the public in fire-protection matters than in British Columbia.[13]

Throughout the BC Forest Service's history, no other activity or concern has received such attention, aside from the gathering of revenue for provincial coffers.

Prior to the Forest Branch's creation, the patrol and lookout system was becoming more efficient, especially in areas of immediate concern. Much of the north would remain left to its own devices, but throughout southern and central British Columbia a growing network was emerging. Even before the Forest Branch was organized, weather was being monitored and the causes of fire and extent of damage was being recorded.

View from Hicks Lookout, 1952. I-71588.

In 1929 two extensive studies were prepared, largely relating to fire protection in the Prince George and Kamloops forest districts. These reports included an inventory of all equipment and also proposed recommendations for additional projects to provide improved protection from wildfire. At this time Prince George was divided into seven ranger districts reporting to the district forester based in Prince George, and C.D. Orchard was the district forester. The ranger districts were staffed by:

		Located in
Ranger District 1:	Ranger: W. Lowry	McBride
	Assistant Ranger	McBride
	Patrolman	Beaver River
	Assistant Ranger	Shere
	Patrolman	Red Pass
Ranger District 2:	Ranger: A. Chisholm	Dome Creek
	Assistant Ranger	Penny
	Patrolman	Goat River
	Lookoutman	Longworth
Ranger District 3:	Ranger: G.M. Kerkhoff	Aleza Lake
	Assistant Ranger	Hutton
	Patrolman	Aleza Lake
	Assistant Ranger	Giscombe
Ranger District 4:	Ranger: G. Forbes	Prince George
	Assistant Ranger	Prince George
	Patrolman	Hixon Creek
	Patrolman	Summit Lake
Ranger District 5:	Ranger: A. Hunter	Prince George
	Assistant Ranger	Isle Pierre
	Patrolman	Mud River
	Lookoutman	Pilot Mountain
Ranger District 6:	Ranger: R.F. Heath	Vanderhoof
	Assistant Ranger	Vanderhoof
	Lookoutman	Tsinkut Mountain
	Assistant Ranger	Fort St James

	Lookoutman	Mount Pope
Ranger District 7:	Ranger: W. Wilson	Fort Fraser
	Dispatcher	Fort Fraser
	Lookoutman	Fraser Mountain
Peace River:	Patrolman	Pouce Coupe
	Patrolman	Hudsons Hope
	Patrolman	Finlay Forks

Kamloops at this time was a much smaller district, as the railway belt was still under the jurisdiction of the federal government. The district forester in 1929 was L.S Hope, and the ranger districts were staffed by:

Ranger District 1:	Ranger: J. Boydell	Blue River
	Assistant Ranger	Blue River
	Assistant Ranger	Valemont
	Patrolman	Valemont
	Patrolman	Thompson Crossing
	Patrolman	Blue River
Ranger District 2:	Ranger: W.R. Noble	Birch Island
	Assistant Ranger	Auldgirth
	Patrolman	Round Top
	Patrolman	Otter Creek
Ranger District 3:	Ranger: G.J. Mayson	Barriere
	Assistant Ranger	Barriere
	Patrolman	Barriere Forks
	Patrolman	Fadear Mountain
	Lookoutman	Baldy Mountain
In Addition:	Assistant Ranger	Adams Lake
	Patrolman	Upper Adams River
	Patrolman	Upper Adams River
	Patrolman	Garnet–Red Rock
	Patrolman	Garnet–Red Rock
	Assistant Ranger	Seymour Arm
	Patrolman	Humamilt Lake

S.E. Marling, with rangers J.K. Frost and A. Chisholm in 1931.
NA-03941.

The transfer of the federal railway belt to provincial jurisdiction in 1930 resulted in formation of an additional five ranger districts (Kamloops, Chase, Salmon Arm, Sicamous and Revelstoke). The Depression of the 1930s, however, meant the progress achieved in the 1920s was suddenly curtailed:

> Considerable progress was made on all fronts until 1931 when government cut-backs significantly reduced the workforce. Each year until the fund was reinstated in 1934 the Chief Forester argued for improvements and funding. In 1934 The Chief Forester pointed out that the essence of good fire-fighting is quick action when a fire is small, and all fieldmen were required to attack all fires at once if found when small.... Larger fires involving heavy expense to fight were considered carefully before expenditures were made. Most of these could be fought only at strategic points to protect specific values.[14]

The chief forester continued to argue for upgraded equipment and more men over the next several years. In 1936 he started a new program and assigned one forester to fire control research, and the Forest Branch began to look at improvements: 10 radios, for example, were purchased and installed in lookouts in 1937, with 40 additional sets in 1938.

A district forester and ranger discussing a fire at Kamloops. 1960. NA-20164.

It was not until after World War II that the Forest Service could renew its emphasis on all aspects of forestry, including protection, as more staff and increased funding became available. The government regularly increased the Forest Protection Fund, and as early as 1946 crews were searching out new locations for fire lookouts. During the 1950s great effort went into incremental improvements to the whole system, but there were some dramatic breakthroughs. These included prefabricated lookout buildings, new pump designs and increased use of aviation and radios. The old systems were not abandoned but rather strengthened, including weather monitoring, the volunteer fire warden program and the use of specialized firefighting crews. Industry was also forcibly involved. Their role became even more carefully defined and monitored. Loggers without the correct equipment on hand were quickly penalized and even shut down.

Assistant ranger Bruce Nobel inspecting a fire guard at the Bill Fire, Blue River, in 1978. NA-34535.

Lookout points were a crucial part of the rapid attack program against fire. Although aircraft patrols were increased for some areas, it was the lookoutman who still reported most fires. The development of a prefabricated design in the 1940s greatly improved the Forest Service's ability to increase the numbers of fire lookouts, and this was greatly assisted by the more extensive use of helicopters after 1952, when, for the first time, successful experiments were undertaken to move the building materials for a lookout at Beaver Mountain in the Prince Rupert district. The parachuting of supplies to remote locations and the use of radios meant expensive telephone lines and trails were not required, and the lookoutman's ability to quickly spread the warning was tremendously improved. If two lookouts in the area could see the fire, the location could be accurately pinpointed on a map by the use of the fire finder. Panoramic photographs were supplied to most lookouts, and the new Macdonald-Bennett fire finder increased the lookout's accuracy in reporting fires and greatly increased firefighters' ability to quickly reach a fire.

In 1951 the large Vancouver forest district operated 27 lookouts manned by 30 men, with 23 of these on Vancouver Island. Among the

Bulldozer creating a firebreak for the Elk Fire (No. 7), 1951. NA-13287.

fires spotted was one on June 21, 1951, when smoke was detected near Campbell River (Elk Fire #7). At its height, about 775 men were engaged fighting the fire: 475 were hired by the BC Forest Service, and an additional 300 loggers joined in. More than 20 bulldozers created firebreaks. The fire covered 23,000 acres (9300 hectares) before being controlled. During the fire, a lookoutman provided a vital service by spotting a small blaze that had jumped the firebreak, and the ranger immediately dispatched the nearest suppression crew, who quickly doused this blaze before it could spread.[15]

A meteorologist was hired by the Forest Service by at least 1948, when D.W. Perrie gave weather forecasting new prominence, although E.E. Gregg, then supervising the Forest Service's protection division, glowingly described in 1945 a ranger who had persisted in using a Sling Psychrometer to measure relative humidity, despite the scoffing of the local loggers at a fire 20 years prior. Gregg credits the ranger, despite his embarrassment, as leading the way to establish the criteria for fire weather determination. By 1957 the BC Forest Service's meteorologist, J.A. Turner, could describe in precise detail what weather factors needed to be monitored to establish a fire danger rating.[16] The main proponent of weather prediction was the Canadian Forest Service, and experiments by J.G. Wright had begun as early as 1925. Based on his and others' work, and in cooperation with the provinces, the Canadian Forest Service developed the Canadian Forest Fire Danger Rating system now used across Canada.

The Forest Service began organizing standby suppression crews shortly after the outbreak of World War II, mainly with high school boys. In 1947 they were paid $75 to $90 a month and were in camp during the fire season. Each camp had a cook and a foreman, and the crew was on standby 24 hours a day. When not fighting fire, the crew worked an eight-hour day doing various jobs, such as clearing trails, stringing telephone lines, cleaning park areas and maintenance and improvement work. The local ranger directed their actions and would determine which fires to send the crew to as a hard-hitting initial suppression activity. As the 1950s progressed, more and more of the crews hired were university students and included women, at first in separate camps. Additional crews were spread around the province, but they remained small in size, usually about 10 in total. In 1950 the five

Using water hoses at the Bill Fire, Blue River, 1978. NA-34511.

suppression crews located in the Vancouver forest district were placed at Langford, Hillcrest for the Cowichan Valley, Nanaimo, Parksville and Quinsam (near Campbell River). The pay rate had risen to $110 a month. By 1955 the protection division employed 16 specialized suppression crews of 10 men each. These were usually located in critical areas where road access was possible, and they were generally deployed as first action response. In 1955 the crews attended 158 fires and held 97 per cent of these to less than five acres (2 hectares).

There still were 1384 fire instances in 1955, mainly caused by lightning. The rangers were assisted in their work by 981 honorary fire wardens who watched for and even took action against fires spotted. The provincial and district fire atlas[17] was up to date, and fire protection planning for each managed forest area was underway. Companies operating forest management licences were obligated to prepare protection plans for their licences.

Rangers also had the power of a police constable in certain instances. They were allowed to arrest someone for refusing to fight fire, for example. It was not uncommon during fire season in the small interior communities for the ranger to round up firefighters from the local pubs and off the streets. It was also not uncommon for local young men to quickly leave for safer climes the moment they heard there was a fire in the area. For a few, it was a way to earn a little extra spending money, but the work of firefighting was hard, dirty and sometimes dangerous.

Bob Perchie: Retired Forest Service Officer

Fighting fire was often easy. In those days managing crews was difficult. One day into the fire you would hear "I have to go home, my wife is going to have a baby. I have to go home because I'm sick." Managing crews with pick-ups is always difficult. This first summer, it was the August long weekend in 1969. Larry Espenant was new. I wasn't on standby. The new assistant ranger was.
He went down to Fort Saint James reserve to get some firefighters. There was nobody around. He could find absolutely nobody.

It was a lovely Sunday afternoon around noon hour. I get a phone call. I go down with him. We go to the church. We go to the basement, and there are 70 or 80 young guys hiding. So we got them. And before they got in the airplane, Brash taught me you always have them drop their bag. They had to hold a bag four feet up and drop it. Because if they had any liquor bottles in it they were going to break. You always had to be kind of careful when you took the crews out. Like I say, managing crews was more difficult than managing fires. It was the long weekend also in August. Most of the crews were out fighting various fires.

We had to go down to the local hotel. It was Saturday and a lot of American tourists were in town for fishing. I phoned the RCMP first and asked for help. They took their bun wagon and parked it in the exit at the beer parlor with the doors open. Then myself and another fellow walked into the café, and you could hear the bar exiting because the Forest Service was looking for firefighters.

Well, we got the firefighters, and this fire was just right out of town, walking distance. A lot of them were not suitable for firefighters. We gave them a pulaski and a shovel, and I can still remember. They could walk up to this fire. I followed along sometime later, and there would be shovels and pulaskis laying along the side of the road along with the odd person. We had a blue slip, a blue book, where you wrote down the guy's name. It was a summons. If he didn't go to the fire, he was going to be in court. Managing the fire was a lot easier than managing personnel.

1958 was the record fire year of the 1950s, and interestingly, the largest numbers were caused by railways (29.7 per cent). Most of these occurred along the Pacific Great Eastern Railway (1086 of 1224 railway-caused fires). The decade's improvements were given a thorough test in 1958 but still proved inadequate. There was increased use of aircraft, including the wider use of water bombing, but most aircraft hours were still dedicated to spotting fires and hauling supplies. Some success was achieved by water bombing hot spots and delaying the spread of lightning strike fires until a ground crew arrived. Helicopters were also proving more useful, and two types of Sikorsky and two types of Bell helicopters were used in 1958, for a total of over 1100 hours. A few helicopter hours were also directed at water bombing. Experiments with helicopters had begun near Prince George as early as 1950, when crews and supplies were quickly transported to a number of fires. Chemical additives were also being tested to retard the spread of fires. Eighteen tons of sodium calcium borate were dropped in

Swiss firestorm area, 1983. ZZ-95152.

1958, but results from tests on the use of additives were inconclusive.

The disastrous fire season of 1958 spurred changes in 1959. The large Vancouver forest district was reorganized in 1959 into 10 fire closure zones so that activities in the district could be regulated with greater precision. Throughout the province the new fire danger signs were being erected at each ranger station so that passersby could quickly determine what the current hazard rating was and act accordingly. The helicopters had proved their worth, when on August 1, with only eight hours of flying time, a big Sikorsky S-58 had moved 50 men and 10 tons of supplies to quickly douse a fire. Don Owen, assistant protection officer for the Vancouver forest district, reported that without the helicopter it would have taken 125 men and two weeks to perhaps get the fire under control.[18] Although water bombing was still far from proven, useful new techniques to make it more effective were being developed. For example, markers were used in some of the 1958 fires to direct the drops, initially using adding machine paper rolls and half rolls of toilet paper. That the half rolls were used resulted in many a comment.

The 1958 fire season also pointed out problems in the current organization. After congratulating the field staff of the BC Forest Service for their "splendid job," Keith Gibson, in the *British Columbia Lumberman,* pointed out that the BC Forest Service crews were "underpaid, understaffed and underequipped." He supported his argument by referring to the

Bell helicopter transporting men and equipment at Mil Fire to Mildred lake, 1960. NA-20265.

Chipmunkrun Creek fire in the Chilliwack ranger district, which burned for 32 days and covered 5000 acres (x hectares). About 120 men were recruited, and shovels and pulaskis were plentiful, but only 14 water pumps were initially on hand and four failed as soon as they were put into service. Five bulldozers were rented, and some air support was available. Four Beavers with 70-gallon (265-litre) tanks flew for five days, and four Avengers with 600-gallon (2270-litre) belly tanks were used off and on for two weeks. A Bell helicopter was eventually available and ferried crews to the fireline, then away when danger threatened. It was supplemented for a short time by the larger Sikorsky S-58. An unidentified ranger "stated flatly, if this 'firetool' [the helicopter] had arrived in time the 'Munk' fire could have been confined to ten acres."[19] The disastrous season resulted in Chief Justice Gordon Sloan being appointed to investigate fire equipment and needs. The next decade brought additional techniques and greater precision to forest firefighting. The BC Forest Service's expertise now is such that it is regularly sought after by other agencies and countries.

At the Interface: Fires Between Town and Forest

The BC Forest Service faced new challenges as the decades progressed, and one emerging critical problem was highlighted as a result of the 1973 Eden fire located near Salmon Arm. Previously, the occasional ranch had been caught within a raging wildfire, but now it was clear that whole communities were potentially at risk and that the government, and especially the Forest Service, needed to do something to mitigate this situation. The Eden fire went out of control on September 11, 1973. It began as an escape from a slash burn and, mainly as a result of high winds, spread rapidly into an uncontrolled blaze that swept down the Gleneden Valley, destroying 18 homes and considerable livestock. The publicity resulting from this fire initiated a process of growing concern over interface fires that would eventually culminate in what became known as Firestorm 2003.[20]

The Eden fire was not the only fire to threaten Salmon Arm. On July 29, 1998, a lighting strike fire, despite immediate response by the BC Forest Service, approached the community. This time more than 40 buildings were destroyed and 7000 people had to be evacuated. In the severe fire season of 2003 an increasing number of interface fires occurred. The *Firestorm 2003* report, initiated by Premier Gordon Campbell, lists 15 communities threatened by fire in 2003. The largest fire was the Okanagan Mountain Park fire near Kelowna.

when 238 homes were destroyed or damaged. The awareness of interface fires that began with the Eden fire and was highlighted by the Mountain Park fire has resulted in an expanding commitment on the part of government to actively work on curbing such destructive fires.

The 1973 Eden Fire in the Okanagan was one of the first major interface fires. NA-29728, NA-29727, NA-29738.

Gordon Comeau: Retired Forest Officer, Princeton

When the scale of the fire was such that they felt that the local district was over their heads and not capable of dealing with it, they would send in what they called an overhead team. It was a six-man team, and each member had a specialty. One person's specialty would be aerial operations. One was the fire boss. He oversaw all of the other five. And a line boss – he oversaw all of the activity on the fire line. There was an admin officer. He made sure payroll, all the paperwork was done according to plan. There was an expediter, which is where I started. And my job was to make sure if they needed something, they got it. If they needed a Cat, I would find them a Cat. If they needed food, I would find the food supplies. If they needed cooks – that kind of stuff. There was an equipment manager. He oversaw all of the equipment. So I think that makes up six in total. And so they specialized in those areas.

That changed a little bit over time, with the different roles and responsibilities. But I acted as an expediter for a couple of years. And then I took a real keen interest in air operations, and I became the air operations boss for a number of years under a couple of different fire bosses. The fire boss just sort of hovered over everybody and over the fire. And he's the one that had the strategic plan as to how we would pinch this off, or direct attack, or indirect attack. And we would meet every evening after and debrief everything, and debrief, and he'd set the goals for the next day. Off we go again.

Punk Fire, Tatla Lake, 1969. NA-33198.

A World-class Operation

In 1994 the provincial government decided that a change was needed in the way protection was managed. Up to this point, regional staff were called from other duties to fight fires, and a decision was made that dedicated crews were required. The Forest Service had been moving toward this approach for some time, with more and more firefighting responsibility shifted away from the rangers to more specialized units. During the next 10 years, refinements resulted in a world-recognized firefighting organization. The crews are divided into three-man initial attack crews, of which 330 men are stationed at about 50 bases around the province. In addition, from the rapattack base at Salmon Arm, the three-man crews can be quickly dispatched to remote areas where, from helicopters, they descend on ropes to reach the fires. The parattack program is based in Fort St John, and those crews parachute from fixed-wing craft to control the wildfire. Twenty-man unit crews are available when a fire gets out of hand and needs greater manpower. Advanced detection and dispatch techniques are also used, including lightning detection, infrared detection and the latest in computer monitoring and crew dispatch.

Avenger making a Firetrol drop with red fire retardant, 1971. NA-33200.

In 2003 the Wildfire Branch faced its most disastrous season to date, and not just due to the hectares of forest burned. Some 334 homes were destroyed, and 50,000 people had to be evacuated. During the worst period, the Forest Service employed 7668 people to fight fire. In addition, 25 air tankers and 202 helicopters fought 880 different fires. As the *Firestorm 2003* report noted:

> Walls of flame moved through the forests at a speed of over 10 metres per minute to threaten Interior businesses. It was the worst wildfire disaster in memory, but why did "the Big One" happen in 2003? The main contributing factors, observers agree, were the hot, dry weather and the resulting high-risk conditions in British Columbia's forests.[21]

Fighting the Good Fight

A second new unit that resulted from the Forest Practices Code was the Compliance and Enforcement Branch. This branch is tasked with ensuring that all parties using the provincial forest obeyed legislation and regulations, and the branch was authorized to enforce these regulations through fines. The branch is semi-independent in that it has its own director responsible for ensuring industry compliance with the guidelines. District and regional enforcement officers reported initially to the district and regional managers, but they also reported on activities to the branch director, who ensured that adequate independence was maintained. In the year 1998–99, for example, the branch undertook 28,644 harvest and road inspections and discovered 1731 problems. Many of these problems related to road construction, such as improper culverts and even road maintenance, while others might deal with damage to Crown forest, failure to use the proper timber marks or improper planting procedures.

The defeat of the NDP government in 2001 and transition to Gordon Campbell's Liberals meant a policy revision and change in Forest Service work arrangements. The central goal of the new Liberal government was to revise the Forest Practices Code, and with this in mind they enacted the legislation currently in effect: The *Forest and Range Practices Act* that came into force January 31, 2004. The new act was described as a results-based approach or a performance approach. The method that the act proposed for regulating British Columbia's forest was described thus: "Government sets objectives and desired outcomes, and forest companies propose results or strategies that reflect these. The companies are then accountable for the results through a rigorous government compliance and enforcement regime."[22] This program still has very mixed reviews, and what can be described as forest health will indicate its success

Planting northeast of Kamloops, 1980. NA-36339.

or failure – although at present the ministry limits this designation to assessing the actual damage to the forest from natural causes, not the overall health of the forest that British Columbians today are seeking.

A Red Forest

The Mountain Pine Beetle outbreak, the focus of so much recent attention, has wreaked havoc on interior forests of British Columbia for centuries, as have other beetles and disease occurrences. At times some of these outbreaks, such as that of the Spruce Beetle, reached epidemic proportions, but it is the Mountain Pine Beetle that first created forests that tint red and then grey as the trees die. The Spruce Beetle has continued to contribute significantly to the problem as well, but the Mountain Pine Beetle, in particular, has been an unstoppable force of nature in which the agencies involved sought to cope in the best way they could. The only solution, though, is the return of cold weather (-20°C in the fall or -40°C in the winter) or, as has happened across central and southern BC, the near complete death of the pine forests. These now dead forests mean that Mountain Pine Beetle populations in BC have declined in recent years, and the end is in sight. But it is expected at least 80 per cent of the Lodgepole Pine forest will be dead by the time the beetle has run its course.

There are various reasons given for the outbreak. Some blame the BC Parks Branch, as the largest initial outbreak began in Tweedsmuir Park. But the beetles were emerging elsewhere, and it was only a matter of time before an uncontrollable outburst of beetle activity occurred in the pine forests. The combination of diligent Forest Service wildfire control, an even-aged forest of mature pine and many mild winters gave the beetles the chance they needed to begin their feast. Outbreaks are not new, but generally in the past, cold weather or naturally occurring fires controlled their spread.

Skidder harvesting timber killed by pine beetles at Dungate Creek, 1975. NA-31060.

The danger of bark beetles was recognized early in British Columbia's forest history. Federal entomologist J.M. Swaine, in a study of insect problems published in 1914 for the BC government, noted that the Mountain Pine Beetle was a serious concern.

> Hundreds of trees have been killed by the beetles since the outbreak commenced about three years ago, and a large number are now infested with the adults and larva. Throughout the whole area affected, the number of killed and dying trees must be very large. The trouble may be noticed from Vernon through to Nicola, a distance of 250 miles. It appears to be widespread throughout most of the Bull Pine area; but a few localities are more seriously affected. The valleys of the Similkameen and Tulameen, in particular, are threatened with serious loss.

He again urged his case in 1917:

> The Bark-beetles constitute the chief enemies of our coniferous forests, and it is impossible to even give an approximate estimate of the enormous loss caused by their depredations throughout Canada. Much of the dead timber whose destruction I attributed to fire is the result of outbreaks of bark-beetles; this is particularly true in British Columbia.[23]

In 1923 Ralph Hopping, head of the federal entomological lab in Vernon, told the Entomological Society of BC that outbreaks of beetles, and especially bark beetles, caused considerable loss. Concerned for the "virulence and spreading nature" of the attacks (especially in Yellow Pine, due to its economic importance), he argued that active control was essential. Lodgepole Pine was even worse hit throughout the 1920s and into the 1930s. In 1923 Hopping noted that "The largest outbreak in Lodgepole pine is in the Yahk district…, but various epidemic areas of greater or less magnitude occur between Penticton and Nelson and between Penticton and Princeton, and east of Okanagan Lake."[24] Ralph's son George later wrote that between 1920 and 1945 there were three major outbreaks of Mountain Pine Beetle in British Columbia, with the largest covering hundreds of square miles in the Tatla Lake area and killing

80 per cent of the trees.[25] Various beetle outbreaks continued throughout the next four decades, until finally the current outbreak began in about 1995.

Beetle attacks were and are a dramatic event. Hector Richmond, surveying beetle outbreaks near Aspen Grove in the early 1930s, recorded witnessing an attack near his camp:

> It was a spectacular sight. The trunk of the tree was swarming with literally thousands of adult beetles, all actively engaged in chewing their way through the bark to the inner wood. The entire swarm had arrived within an hour. By dawn the following morning, the entire population had penetrated the bark and were no longer visible.[26]

Such attacks did not even come close to capturing the destruction caused by the clouds of beetles that descended during the peak of the most recent outbreak.

Trees killed by Mountain Pine Beetles in the Suskwa region, 1975. NA-31051.

The BC Forest Service was very much concerned with other insect pests as well. In 1955, for example, the Forest Service was more concerned with outbreaks of the Blackheaded Budworm, the Spruce Budworm and the Douglas-fir Beetle. The Mountain Pine Beetle, however, was a target of great concern and the one to watch. In 1981 a BC Forest Service brochure called *The Tree Killer*, noted a large-scale outbreak in the Nelson district. At that time, despite decades of experimentation in the United States, it was believed that the only way to control an outbreak was to log an area around the affected area to prevent the beetle from reaching new trees.[27] This later proved a hopeless task, as the beetle could travel hundreds of miles.

The current outbreak, described as starting in 1995, is actually part of an ongoing saga that has grown in severity. During the 1960s fir and spruce beetles had been the most significant insect tree killer, but by the 1970s the Mountain Pine Beetle was on its upward trend.[28] A federal report noted in 1982 that "in the Cariboo Region, the widespread infestations showed no signs of abating, except where most mature pine have been killed...."[29] The BC government, in cooperation with the federal government, instituted a beetle management program, and considerable effort was directed to controlling the beetles' impact on the forests and the communities that utilized this resource. In 2010 a Forest Service report estimated that about 51 per cent of the total merchantable pine volume had already been killed and that by 2016 this will have increased to about 59 per cent. A decline in the beetle population is then expected to occur.[30] This seems to be the case, as the number of attacks throughout British Columbia has recently decreased, but also much of our Lodgepole Pine forests no longer exist.

The Forest Service divided the province into Emergency Bark Beetle Management Areas, and only several northern areas are now being aggressively tackled. Most of the rest of the province is undergoing salvage harvesting, and consideration and study is being given to forest restoration. Active community programs have also been implemented to try and mitigate the impact of beetle outbreaks on the affected communities.

The Small Business Forest Enterprise Program, created in the early 1980s in response to the Pearse report and expanded in the 1990s, was renamed BC Timber Sales on April 1, 2003, and established as an autonomous program within the ministry. The new legislation introduced in the mid 2000s has pulled

30 per cent of the allowable cut away from the major licensees, and 20 per cent of that cut is allocated to BC Timber Sales, while eight per cent is allocated for First Nations and two per cent for community forests. All of these programs are intended to spread the forest allocation to a wider variety of interested parties. BC Timber Sales is expected to provide smaller loggers with a better chance at making a living. Forestry costs of the major licensees has, by and large, been turned over to the licensee, while the smaller operators are unable to cope with reforestation, and the BC Forest Service continues to provide them these services. BC Timber Sales, in effect, acts like a major licensee; it advertises timber for sale and then allocates the timber to highest bidder. BC Timber Sales prepares silvicultural prescriptions and engages in reforestation tasks. The tree nurseries still operated by the ministry are now under the jurisdiction of BC Timber Sales and largely supply the needs for these smaller loggers and their reforestation requirements. A review in 2006 recommended that BC Timber Sales remain part of the Ministry of Forests.

The First Nations allocation emerged from two court cases that determined the Crown had an obligation to accommodate potential aboriginal rights. The decision created huge uncertainty. In an effort to mitigate this uncertainty regarding timber, and following considerable consultation, the Forest Service adopted a strategy of negotiating with interested First Nations where rights and title might be affected. Three central polices were implemented: revenue sharing, direct award timber sales and Forestry and Range Agreements. Since being introduced in 2003, over 128 such Forestry and Range Agreements have been signed and most remain in effect. The exact details of such agreements vary considerably and generally range from acknowledgement of rights to revenue sharing or assistance with problems, such as reforestation. For example, the Gitanyow agreement signed August 3, 2006, allows the Gitanyow to apply for a non-replaceable timber licence in the Cranberry Timber Supply Area of 86,000 cubic meters annually for five years. The actual mechanism of transfer of timber and harvesting rights has happened in variety of ways. TFL 38, for example, was sold to the Squamish First Nation in 2005 by International Forest Products. Under the Forest Revitalization Act, the company was going to lose a large portion of the cut from this area and decided that the sale of the whole licence was in the best interests of all parties concerned. Several successful First Nations forest operations were initiated prior to the 2003 legislation, including the innovative model established by Iisaak Forest Resources Ltd. on July 24, 1998, which is owned by the Central Region Nuu-chah-nulth First Nations.

The community forest program was directed at providing local communities an opportunity to have more say in how nearby resources are managed.

Although it was an amendment to the *Forest Act* in 1998 that created this form of tenure, several communities had already been engaged in forestry. The district of Mission began operation of a tree farm licence in 1958, while the district of North Cowichan has managed forest land since 1946. The North Cowichan lands were converted into woodlots in the 1960s but are now managed as a community forest. The third experiment with community forestry prior to the 1998 legislation was in Revelstoke, where, after several years of sawmill closures and negotiations and involvement in planning activities, the city of Revelstoke was granted TFL 55. Following the introduction of the 1998 legislation and by the end of 2004, 16 additional community forests were in place throughout BC, and in 2009 they were removed from their five-year probationary status and granted 25-year licences.

New programs continued to come regularly under the Liberal government. In 2003 an endeavor initiated in the Forest and Range Evaluation Program, under the direction of the chief forester, was to evaluate the effectiveness of a results-based approach to forest management in British Columbia. It examines a wide range of activities that impact forest biodiversity and quality. In his 2010 report Chief Forester Jim Snetsinger noted eight values are currently being monitored: fish/riparian, water quality, stand-level biodiversity, visual quality, cultural heritage, timber, forage and soil. In each category a central question is asked and answered. The question for visual quality, for example, asks: "How well are we managing and conserving views in designated scenic areas? and are established Visual Quality Objectives being achieved?"[31] In 2005 the Forests for Tomorrow program was introduced and the Future Forest Ecosystems Initiative was announced. Forests for Tomorrow is intended to provide a means to combat the massive wildfires of the early 2000s and the catastrophe brought on by the Mountain Pine Beetle outbreak. The program's primary initiative is reforestation, first in the burned areas and then in the insect infestation areas. The Future Forest Ecosystems Initiative was established to provide guidance on climate change and to understand how our forests will change in the future.

The BC Forest Service has undergone more change in the last 20 years than throughout its previous 80 years. It has changed ministries more often, been reorganized more often and implemented more significant legislative changes than ever before. The 1990s and beyond have been times of great struggle in British Columbia's forests, and how they should be managed has come under scrutiny and been subject of debate and dissent. Downward adjustments of the allowable annual cut had begun as early as 1980, and this trend continued into the 1990s, as the chief forester began again reducing the allowable cut in several districts. In 2003 the then Chief Forester Larry Pedersen told the University of

British Columbia Faculty of Forestry that he was optimistic for the long-term future of BC's forests, but

> Managing our way through the transition from the preponderance of mature, naturally grown forests to the managed forests of the future will be a continuing challenge. How our society chooses to use or allocate its valuable forest resources is likely to be reviewed, evaluated and debated in each successive future generation. While our current approach to sustainability attempts to meet obligations to future generations, the one thing that we cannot do today with any certainty is predict what the preferences of future generations will be, given the ever changing context of local and global affairs.[32]

Dirk Trigg

Well, I'm hoping now that we are one big team. In the southern interior forest region we are doing things the same. But there's no question that there was differences in ... you know legislation. You can't go against the law and the regulations, but policy is policy. And policy can be interpreted in different ways.... There's been a culture in the ministry of problem-solvers, whether we think we've got the tools or not. If there's a problem out there, perhaps a licensee can't figure out how to go into a drainage to harvest timber or someone just can't get enough volume through the competitive process or whatever, the ministry has always had a history and a culture of trying to resolve these issues and come up with options or whatever. And so the different regions have tried to be innovative. Basically, that's what we were doing: how do you solve these problems? We're all trying to manage the forests in the best manner, create a world competitive industry, make sure there's lots of small players and big, quite a diversity. And so, yes. But you can see as being a leader in trying to tie the old three regions together, when you're chatting with the staff that came from each of these directions trying to build your new policy, your new procedures, you can see the different flavour on how they would have solved the same problem you would have. And it's interesting how things have evolved. Basically, there was more isolation, right? You dealt with things in your district. If you couldn't deal within the district, you got help from the region. So now we're in a bigger world where the districts all come to one region in the centre here. So we're a lot more standardized and are a lot more on the same page.

Communications centre, Kamloops, 2012. IMGP3852.

But yeah, there was cultures of solving things and being innovative, and you might not hear about how another region was being innovative until you went to Victoria for a provincial meeting where all the regional staff would get together and you'd hear about it. But now, you hear about things a lot faster. It's a faster moving, closer world just because of technology and conference calls and all that stuff. Whereas we didn't do that much in the early years, right? A conference call was expensive. You might do one of those every two months between the regions or whatever. Now it's a daily thing to have three or four conference calls and chat with people anywhere in the province. It's a different world – and email and the whole bit. Everybody's communicating immediately, whereas when I was in the Horsefly forest district, I didn't send a single email, right? And I could go back to that world, I'll tell you.

The government of British Columbia has implemented a variety of Forest Service-related activities in recent years, seeking to meet its objectives and obtain advice on how to administer the forest. In 2008 Rich Coleman, the Minister of Forest and Range, initiated a forestry roundtable, whose report was published in early 2009.[33] Follow-up to the roundtable's recommendations has extensively revamped the way BC's forests are managed, ranging from revised tenures and licensing to new marketing tools, working with First Nations and support for bioenergy projects.[34] Since that time, one of the ministry's goals has been to implement the roundtable's 29 proposed recommendations. Part of the process of streamlining administration of provincial resources meant the Forest Service, along with a variety of other government operations, was merged into the Ministry of Forests, Lands and Natural Resource Operations. As the ministry's recently revised service plan suggests, the objective is:

> The "one project, one process" approach for major investments creates a single framework that is timely, diligent and science-based. Now, instead of going to nine different agencies to apply for over 1200 different types of provincial authorizations, there is one ministry to coordinate and streamline these operations.[35]

This fulfills a key recommendation, perhaps most publicly known through the Peel Commission report of 1991, when the Chair recommended that "the Government of British Columbia establish a new Ministry of Renewable Natural Resources to replace the Ministry of Forests and take on an expanded role in the policy, regulation and general administration of the forest and rangelands of the province." This new ministry would be responsible for "policies and regulations related to the key values of timber, water management, range, recreation, and fish and wildlife. The land base managed by the Ministry will be those lands not allocated to the Forest Resources Corporation."[36]

The most recent legislative and operational revisions and the creation of the large superministry move the Forest Service dramatically beyond the 1991 recommendation, and only time will reveal how successful it will be. The BC Forest Service has faced monumental challenges in the past, and this historic and proud organization has undergone many changes in character and even in direction, as have been partially charted in this volume. But through it all, the Forest Service has survived and become stronger, and it continues to fulfill its mandate of managing, protecting and preserving BC's forests for the benefit of all British Columbians.

Epilogue

We were in the interior on our way to conduct staff interviews the day Premier Campbell made his sudden announcement to end the BC Forest Service and divide its duties among two ministries.[1] Did the BC Forest Service still exist? We were uncertain what we would find at the regional office in Kamloops. We found shock, a degree of confusion and staff meetings underway. Most striking was the tremendous depth of self-identification and tangible grief among the dedicated staff for the proud heritage of the BC Forest Service. Yet the staff made time for us on one of their worst days and shared some of the most heartfelt interviews we heard.

We then discussed our own feelings. Did we want to continue? How should we deal with this as historians? Now, with some distance from that event, the emotional impact of the many changes we noted in the history of the Forest Service and Forest Branch take on a deeper and more personal meaning. Older retired staff members had tried to explain to us about change and realignment, about coming back into step with society and about internal and external criticism. For us these issues now became raw and real in a way few archives can explain.

Looking back, it is impossible to find a decade without such days of shattering change in the Forest Service. What stands out is the workers' depth of pride in their past and in what has been achieved, much of it anchored in communities and conducted one staff person at a time – with a pulaski and a hose, with permits and countertop conversations, in First Nations community consultations, in visits to tree-planting camps or through frank discussions in company offices. All of this work has been done with a tremendous sense that as a whole, each action throughout the good days and bad days (such as the death of a colleague), plus all the countless other duties as required, formed something of tremendous value.

If there is a threat to the core of the BC Forest Service, whatever its formulation, it may not lie with logos and organizational charts or political decisions. It lies with its best new partner: digital technologies that gather and organize vast amounts of data and allow for rapid decisions. The new tools have a cost, limiting the time staff spend physically out in the forests and on the grazing lands, talking and connecting with individuals and the communities they serve. In our interviews we saw staff constantly trying to balance field work and new technologies. Yet there has never been a time without such change.

In our research for this book, we listened to many people describe how their careers began. Time and again, they talked about coming to the Forest Service as young women and men and finding a place on the land doing duties as required: working on trails or surveys, or fighting fires in their first summer job with the BC Forest Service. A few, not all, found something that grew to become a career. Those options for career entry have shrunk drastically today. Staff interviewed were painfully aware of this and expressed concern about succession – not of their own individual job category or section but of the ideals and esprit-de-corps that they believe is taught out on the land, where the ideas behind the BC Forest Service are rooted. Even without the Forest Service name, those ideals are of value to this province.

As we prepare to go to press, Canada's newspapers are filled with discussion about the meaning of the recent Supreme Court of Canada ruling in the case *Tsilhqot'in Nation v. British Columbia.* Originally brought by Chief Roger William, the monumental case is now finally resolved, after 20 years. The Supreme Court ruled that Aboriginal title is unextinguished on non-treaty land and that consultation to obtain permission will be needed in key areas of forestry and other resource uses.[2] A great deal of media attention has been given to the impact of this decision, how it changes everything and will shape the future of forest lands and traditional lands in British Columbia. All true. In the midst of this media storm, a small item also appeared which drew almost no public discussion. On July 11, 2014, British Columbia signed its 150th revenue-sharing agreement with a First Nations government under the new Forest Consultation and Revenue Sharing Agreement.[3] While the Forest Service name did not appear, its staff expertise helped form that announcement. The forests and grasslands of British Columbia remain, and women and men continue to dedicate themselves to understanding and managing our forests. Climate change, water shortages, lightning strikes and careless fires all repeat that annual cycle that focus us as a society to watch, discuss and protect our forests.

To dedicate one's life, and at times to risk one's life for a forest or a range, is something not well understood by the general public. The interface between nature and society is an unusual place to work. That relationship has never been static. Society, the Forest Service, the landscape itself, forests and range, are constantly reshaped, reformed and ever changing. Responding to those changes has at times come too quickly for some, and too slowly for others. But throughout that process there is something quite wonderful in the determination and depth to which these public servants have dedicated themselves to serve both the people of British Columbia and their forests.

Appendix 1

Ministers of the BC Forest Service 1910–Present

William R. Ross (Chief Commissioner of Lands)	Oct. 10, 1910 – Dec. 15, 1915
William R. Ross (Minister of Lands)	Dec. 15, 1915 – Nov. 30, 1916
Thomas D. Pattullo (Minister of Lands)	Nov. 30, 1916 – Aug. 20, 1928
Frederick Burden (Minister of Lands)	Aug. 21, 1928 – Oct. 29, 1930
Nelson S. Lougheed (Minister of Lands)	Oct. 29, 1930 – June 1, 1933
Canon Joshua Hinchliffe (Minister of Lands)	June 1, 1933 – Nov. 15, 1933
A. Wells Gray (Minister of Lands)	Nov. 15, 1933 – May 15, 1944
John Hart (Premier and Minister of Lands)	May 15, 1944 – Nov. 8, 1944
Edward T. Kenney (Minister of Lands until 1945 then Lands and Forests)	Nov. 8, 1944 – Aug. 1, 1952
Robert E. Sommers	Aug. 1, 1952 – Feb. 28, 1956
Ray Williston	Feb. 28, 1956 – Sept. 15, 1972
Robert Williams	Sept. 15, 1972 – Dec. 22, 1975
Tom Waterland	Dec. 22, 1975 – Jan. 17, 1986
Don Phillips (acting)	Jan. 17, 1986 – Feb. 11, 1986
Jack Heinrich	Feb. 11, 1986 – Aug. 11, 1986
Jack Kempf	Aug. 14, 1986 – March 6, 1987
John Savage (acting)	March 6, 1987 – March 31, 1987
David Parker	March 31, 1987 – Nov. 1, 1989
Claude Richmond	Nov. 1, 1989 – Nov. 5, 1991
Dan Miller	Nov. 5, 1991 – Sept. 17, 1992
Art Charbonneau	Sept. 17, 1992 – Dec. 16, 1992
Dan Miller	Dec. 16, 1992 – Sept. 15, 1993
Andrew Petter	Sept. 15, 1993 – Feb. 22, 1996
Denis Striefel	Feb. 28, 1996 – June 17, 1996
David Zirnhelt	June 17, 1996 – Feb. 29, 2000
Jim Doyle	Feb. 29, 2000 – Nov. 1, 2000
Gordon Wilson	Nov. 1, 2000 – June 5, 2001
Michael de Jong	June 5, 2001 – June 16, 2005
Rich Coleman (Ministry of Forests and Range)	June 16, 2005 – June 23, 2008
Pat Bell	June 23, 2008 – March 14, 2011
Steve Thomson	March 14, 2011–

Appendix 2

Ministers of the BC Forest Service 1910–Present

H. Cathcart	1929 – 1944
C.D. Orchard	1945 – 1957
R.G. McKee	1958 – 1964
F.S. McKinnon	1965 – 1968
J.S. Stokes	1969 – 1976
E.L. Young (Chief Forester & Chief Executive Officer)	1977
Mike Apsey	June 1978 – June 1984
R.W. Robbins (acting)	1984
A.C. Macpherson	1985 – 1986
R. Flitton	Aug. 1986 – April 1, 1987
R. Marr	1987 – Nov. 30, 1989
Philip Halkett	Jan. 1, 1990 – April 15, 1991
Bob Plecas	April 15, 1991 – Nov. 5, 1991
Philip Halkett	Nov. 5, 1991 – Sept. 15, 1993
Gerry Armstrong	Sept. 15, 1993 – Feb. 3, 1997
John Allan	Feb. 3, 1997 – Sept. 4, 1999
Lee Doney	Sept. 4, 1999 – June 5, 2001
Don Wright	June 5, 2001 – June 29, 2003
Doug Konkin	June 30, 2003 – Jan. 20, 2009
John Dyble	Jan. 20, 2009 – June 1, 2009
Dana Hayden	June 1, 2009 – March 14, 2011
Doug Konkin	March 14, 2011 – April 19, 2013
David Peterson (acting)	April 19, 2013 – October 9, 2013
Tim Sheldan	October 9, 2013 –

Appendix 3

Chief Foresters of the BC Forest Service

H.R. MacMillan	1912 – 1915
M.A. Grainger	1915 – 1920
P.Z. Caverhill	1920 – 1935
E.C. Manning	1935 – 1941
C.D. Orchard	1941 – 1958
R.G. McKee	1958 – 1959
F.S. McKinnon	1959 – 1965
L.F. Swannell	1965 – 1972
I.T. Cameron	1972 – 1974
E.L. Young	1974 – 1978
W. Young	1978 – 1984
R. Robbins	1984 – 1985
John Cuthbert	1985 – Aug. 30, 1994
Larry Pedersen	Sept. 1, 1994 – Nov. 2004
Jim Snetsinger	Nov. 8, 2004 – Mar. 2012
David Peterson	May 10, 2012 –

Appendix 4

Names of the BC Departments and Ministries Responsible for the Forest Service, 1912–Present

Department of Lands, Forest Branch	Feb. 1912 – April 5, 1945
Department of Lands and Forests, Forest Service	April 5, 1945 – March 30, 1962
Department of Lands, Forests and Water Resources, Forest Service	March 30, 1962 – Dec. 11, 1975
Ministry of Forests, Forest Service	Dec. 11, 1975 – Nov. 6, 1986
Ministry of Forests and Lands, Forest Service	Nov. 6, 1986 – July 6, 1988
Ministry of Forests, Forest Service	July 6, 1988 – June 16, 2005
Ministry of Forests and Range, Forest Service	June 16, 2005 – October 25, 2010
Ministry of Forests, Mines and Lands	October 25, 2010– March 14, 2011
Ministry of Forests, Lands and Natural Resource Operations	March 14, 2011–

Endnotes

Introduction

1 Forestry has global ancient roots: forest regulations existed in early Roman and Greek law; and the medieval British Crown managed a diversity of user rights to the forest commons. Forests have always been political – Karl Marx's first publication was on forest policy and firewood rights. Distinct German and French styles of forestry emerged and dominated 19th century thought. The modern profession of forestry in North America emerged with the founding of dedicated forestry schools, such as the influential Yale Forest School (1900) and Canada's first school of forestry at the University of Toronto (1907). Others followed in Canada at the University of New Brunswick (1908) and the University of British Columbia (1920).

2 Cook's crew did make repairs to masts in Nootka Sound in 1778, but John Meares was the first to set out to locate and export spars to offshore markets. W. Kaye Lamb, "Early Lumbering on Vancouver Island, Part I: 1844–1855," *British Columbia Historical Quarterly* 2:1 (January 1938): 31–32.

3 For the Hudson's Bay Company export trade see Richard Mackie, *Trading Beyond the Mountains*. Vancouver: UBC Press, 1997. For Stamp and the early entrepreneurs see James Morton, *The Enterprising Mr. Moody, the Bumptious Captain Stamp*. Vancouver: J.J. Douglas, 1977.

4 William Eddy Banfield to Colonial Secretary, July 20, 1860 BC Archives, F107/3.

5 Lorne Hammond. "Basin of Beads: The Life and Death of William Eddy Banfield." Unpublished manuscript, 1989. Barry Gough. *Gunboat Frontier: British Maritime Authority and Northwest Coast Indians, 1846–90*. Vancouver: UBC Press, 1984. We do not know if Stamp's Alberni loggers were European, First Nations or both. Later years brought conflict and death in incidents ranging from a dispute over a potato patch, to the shelling of an Ahousat village.

6 Robert A.J. McDonald. *Making Vancouver, 1863-1913*. Vancouver: UBC Press, 1996, p. 9.

7 Gordon Hak. *Turning Dollars into Trees: The British Columbia Coastal Lumber Industry, 1858-1913*. Toronto: University of Toronto Press, 2000, pp. 136-137. See also Richard A. Rajala, *Clearcutting the Pacific Rain Forest: Production, Science, and Regulation*. Vancouver: UBC Press, 1998.

8 Hak, *Turning Trees into Dollars*, pp. 153 - 154. Several of the non-First Nations stevedores were either Kanakas (Hawaiians) or married to First Nations women.

9 Hak, *Turning Trees into Dollars*, p. 137.

10 Robert E. Cail. *Land, Man, and the Law: The Disposal of Crown Lands in British Columbia, 1871 - 1913*. Vancouver: UBC Press, 1974, p. 93. Those familiar with the work know it to be a brilliant M.A. thesis, completed in 1954, but it does include errors. There exists a corrected annotated copy at the BC Archives in Victoria. Cail died in 1958 while completing doctoral studies. See the foreword by Margaret Ormsby, p. xi.

11 Hak lists timber fires, many based around logging operations or sawmills, for 1868, 1870, 1872, 1875, 1882, 1883, 1885, 1887, 1889 and 1891. As steam donkeys and locomotives became more commonly used, fires increased. Hak, *Turning Trees into Dollars*, pp. 76–77.

12 Hak, *Turning Trees into Dollars*, p. 75 and correspondence listed in fn. 46 & 47.

13 Thomas R. Roach, "Stewards of the People's Wealth: The Founding of British Columbia's Forest Branch," *Journal of Forest History*, January 1984: 14 - 23.

14 R. Peter Gillis and Thomas R. Roach. *Lost Initiatives: Canada's Forests Industries, Forest Policy and Forest Conservation*. New York: Greenwood Press, p. 134.

Chapter 1

1 R. Peter Gillis and Thomas R. Roach. *Lost Initiatives: Canada's Forests Industries, Forest Policy and Forest Conservation*. New York: Greenwood Press, p. 138-140.

2 *Final Report of the Royal Commission on Timber and Forestry, 1909-1910*. Victoria: King's Printer, 1910..

3 Robert Howard Marris. *"Pretty Sleek and Fat": The Genesis of Forest Policy in British Columbia, 1903 – 1914.* Masters of Arts thesis, University of British Columbia, 1979, pp.46–47.

4 Transcript of speech of Gifford Pinchot, *Transcript of Hearings. Royal Commission on Timber and Forestry, 1909-1910.* B.C. Archives, GR271, Box 1, File 3.

5 A. Leitch (p. 801). Prohibition of fire from July to September was supported by F.W. McLarne, p. 1004). Supporters for a government permit system included J. Lemay, John Oliver, and D.V. Mott, p. 920. *Transcript of Hearings. Royal Commission on Timber and Forestry, 1909.* B.C. Archives, GR271, Box 2, file 5 and 6.

6 A. Leitch (p. 801). Prohibition of fire from July to September was supported by F.W. McLarne, p. 1004). Supporters for a government permit system included J. Lemay, John Oliver, and D.V. Mott, p. 920. *Transcript of Hearings. Royal Commission on Timber and Forestry, 1909.* B.C. Archives, GR271, Box 2, file 5 and 6.

7 Testimony of Hon. John Oliver, M.L.A. for Delta. *Royal Commission on Timber and Forestry, 1909.* B.C. Archives, GR271, Box 2, file 7, pp. 1191-1192.

8 John Oliver, GR271, Box 2, file 7, p. 1191.

9 John Oliver, GR271, Box 2, file 7, p. 1197.

10 John Oliver, GR271, Box 2, file 7, p. 1203.

11 William Roderick Ross, a Conservative, was elected from Fernie 1903, 1907, 1909, 1910, 1912, and then Fort George 1916. His town was burned to the ground in four hours on April 29, 1904 and quickly rebuilt in brick.

12 "British Columbia's Forest Policy," speech by the Hon. William R. Ross, Minister of Lands on the Second Reading of the Forest Bill, Legislative Session of 1912, p. 24.

13 Ross, "Speech," p. 11.

14 Ross, "Speech," p. 12.

15 James G. Lewis. *The Forest Service and the Greatest Good: A Centennial History.* Durham, North Carolina: Forest History Society, 2005, pp. 46-47.

16 H.N. Whitford and R.D. Craig. *Forests of British Columbia.* Ottawa: Canada. Commission of Conservation, 1918, p.115. The authors note that they drew directly on a memorandum prepared by M.A. Grainger, Chief Forester of British Columbia.

17 Harry Whitaker. "Mathew Foley and the BC Log Scale," *British Columbia Lumberman,* Vol. 65, No. March. (1981), pp. 20–23–29.

18 Report on Fire Protection. *Report of the Minister of Lands,* Victoria: King's Printer, 1912, G23.

Chapter 2

1 H.R. MacMillan, cited in Robert D. MacRae. *The Man Who Gave Back: The Rule of Life of H.R. MacMillan.* Vancouver: School of Theology, 2007, p. 48; citing Donald McKay. *Empire of Wood: The MacMillan Bloedel Story.* Seattle: University of Washington Press, 1982, p. 35.

2 Chief Forester H.R. MacMillan., *Report of the Forest Branch of the Department of Lands, 1912,* p. 25.

3 *Ibid.,* p. 7.

4 *Ibid.,* p. 36.

5 Chief Forester H.R. MacMillan. *Report of the Forest Branch, 1912.* Victoria: King's Printer, 1913, p. 2.

6 Chief Forester MacMillan. *Report of the Forest Branch, 1912,* p. 27.

7 *Forest Branch Report 1914,* p. D100.

8 *Ibid.,* pp. 94-95.

9 Report of Professor J.M. Swaine. "Forest-Insect Damage," *Report of the Forest Branch, 1914,* D. 65–6.

10 *Ibidem.*

11 *Report of the Forest Branch, 1914.*

12 *Report of the Forest Branch, 1917.*

13 Interview transcript, R.V. Stuart and C.D. Orchard. BCA T1871:1-2, Disc 96-A pp. 23–24.

14 Peter Murray. *Home from the Hill: Three Gentlemen Adventurers.* Victoria: Horsdal and Schubart,1994, p.175

15 Interview transcript, R.V. Stuart and C.D. Orchard. BCA T1871:1-2, Disc 96-A, p. 23.

Chapter 3

1 Department of Lands. *Report of the Forest Branch, 1920,* H18.

2 Department of Lands. *Report of the Forest Branch, 1921*, H36.

3 *Forest Branch Newsletter,* No. 28 May 1 1939, 7.

4 Department of Lands. *Report of the Forest Branch, 1937*, M14,and, Department of Lands. *Report of the Forest, 1938*, N22.

5 Author's Interview Transcript, R.V. Stuart and C.D. Orchard. *BCA T1871*:1-2, Disc 96-A pp. 28.

6 *Forest Branch Newsletter,* No. 20 October 18, 1937, 1-5.

7 John Parminter and Ralph Schmidt. *An Early History of the Research Branch,* British Columbia, Ministry of Forests and Range. 2006.

8 "Last Tribute to Forester," *Victoria Daily Times*, 12 December, 1935, 15.

9 Department of Lands. *Report of the Forest Branch, 1937*, M8.

10 Department of Lands. *Report of the Forest Branch, 1937,* p. M 10.

11 Helen Manning Akrigg. *Ernest C. Manning, 1890-1941* (Ministry of Forests Library)

12 Rajala. *Clearcutting the Pacific Rainforest*, 166.

13 Grainger quoted in Akrigg. *Ernest C. Manning*, 4.

14 Manning quoted in Rajala. *Clearcutting*, 160.

15 Rajala. *Clearcutting*, 162.

16 Rajala. *Clearcutting*, 162-163.

17 Akrigg. *Ernest C. Manning,* 7.

18 Akrigg. *Ernest C. Manning,* 6.

19 Fred Mulholland. "B.C. Forest Surveys," *Forestry Chronicle* (September 1931): 185, cited in John Parminter, "From Timber to Biodiversity: The Evolution of British Columbia's Forest Inventory," *Forest History Today,* 2000: 16.

20 Department of Lands. *Report of the Forest Branch, 1936,* p. L 7.

21 *Forest Branch Newsletter* No. 1 May 27 1935, 2.

22 *Forest Branch Newsletter* No. 16 April 1 1937, 5.

23 *Forest Branch Newsletter* No. 12 Sept. 15 1936, 5.

24 This section draws extensively from the Ministry of Forests, Bloedel Fire, "O" series Bloedel Fire, provided to the author in a digital scan. It includes correspondence, reports by staff and the chronology is based upon the Fire Diary of D.B. Taylor. His diary became important in the litigation that followed. A summer fire was followed by four years of litigation over responsibility and costs.

25 "Vast BC Timber Area is Menaced by Forest Fires," *Province*, July 15, 1938. Ministry of Forests, Bloedel Fire "O" Series, newspaper clipping 19.

26 Herald Henry Caulfield "Torchy" Anderson (1894-1984). Victoria born, he was a reporter for the *Calgary Herald* until his military service in World War I. In 1928 he moved to Vancouver where he worked as a political reporter for the *Vancouver Province*, becoming Editor in 1946. He retired in 1955.

27 "Miracle Saves Fortune in Logs From Maw of Advancing Flames", *Daily Province*, July 22 1938. Ministry of Forests Bloedel Fire, "O" Series, newspaper clipping 34.

28 Ministry of Forests. Ministry of Forests, Bloedel Fire, "O" series, Taylor Fire Diary.

29 *Forest Branch Newsletter*, No. 26, November 20, 1938, 1.

30 Department of Lands. *Report of the Forest Branch 1942* FF51.

31 *Forest Branch Newsletter*, No. 26, May 1 1939, 6.

32 *Forest Branch Newsletter*, No. 17 May 29, 1937, 4.

33 Department of Lands. *Report of the Forest Branch 1939*, E7-E8.

34 Haig-Brown, quoted in Akrigg. *Ernest C. Manning*, 12.

35 Department of Lands. *Report of the Forest Branch 1943* BB46.

Chapter 4

1 "Are We Overcutting?," *Truck Logger*, (July 1951), 6.

2 "Forest Management of Paramount Importance to Every B.C. Citizen," *British Columbia Lumberman*, (April, 1948), 56.

3 C. D. Orchard to District Foresters and Heads of Divisions, February 4, 1949, circular letter 2079.

4 From *Forest Service Newsletter*, #105, April, 1952.

5 Fulton, Royal Commission D14.

6 Forests, Annual report 1913 d 76.

7 D.L. McMullan, "The work of the Surveys Division – British Columbia Forest Service." Thesis, Department of Forestry University of British Columbia, 1932.

8 "Orchard Defends Govt. Policy," *The Truck Logger*, (February, 1952) 36.

9 "Are We Overcutting?," *Truck Logger*, (July 1951) 30.

10 E.W. Bassett, District Forester to Chief Forester, December 7th, 1950, Ministry of Forests Lands and Natural Resource Operations, Public working Circles, file 0187119-1.

11 F.S. McKinnon to C D Orchard July 27th, 1950, Ministry of Forests Lands and Natural Resource Operations, Public working Circles, file 0187119-1.

12 Ibid Dec 13, 1951.

13 F.S. McKinnon to THE CHIEF FORESTER, July 18, 1951. Ministry of Forest, filed 0189897.

14 Officially, working plans division was merged into inventory division on May 3, 1966.

15 E.W. Bassett to all District Foresters, July 28th, 1947, Circular Latter no. 1977, BCFS file 01546.

Chapter 5

1 Larry S. Eckardt "What Tree Farm Licences Have Done To the Forest Industry of B C!" *British Columbia Lumberman*, January 1961, p. 25.

2 R. A. Waldie to H.G. Bancroft, April 4, 1960, MOF File 0187119-5.

3 R.W. Robbins to District Forester Prince George, July 16, 1962, Forests Lands and natural resource Operations, Naver Public Sustained Yield Unit, File 0189897, Section 2.

4 When the parks division moved to its new ministry, 42 permanent staff transferred.

5 D.W. Ruhl to R. G Williston, November 4, 1969, Ministry of Natural Resource Operations, Nechako Supervision, File 0272555.

6 F S Perry, "Change of Pace Impresses Visitors, *British Columbia Lumberman*, (October, 1957), 42.

7 Plus tree: A phenotype judged (but not proved by test) to be unusually superior in some quality or quantity – e.g., exceptional growth rate, desirable growth habit, high wood density, exceptional apparent resistance to disease and insect attack or to other adverse environmental factors.

8 Cowan had previously been a longtime employee of the BC Forest Branch, mainly working in Protection.

9 "Fire Protection in northern B.C." *The Truck Logger* (April 1965), 44.

10 Robert Forbes, "Frontiers of the Sixties," *British Columbia Lumberman* (January, 1961), 13.

11 Pendray started his career in 1937 and then, as many others, went to war. On his return he was posted to Kamloops in 1945 and then moved to Victoria in 1948.

12 J.E. Milroy, "Nelson Forest District Grazing Annual Report, 1956" Ministry of Forest Library, 1.

13 J. E. Milroy, "Nelson Forest District Grazing Annual Report, 1957," Ministry of Forest Library, pp. 2 and 3.

14 M.G. Isenor to Chief Forester, October 17, 1973, Narcosli Planning file, # 0272554, NRO.

Chapter 6

1 See Henry J. Benskin, "Financial Implications and some costs and Benefits of Logging Guidelines in the Chilliwack Provincial Forest," Master of Forestry, U.B.C. 1973. Available online at https://circle.ubc.ca/handle/2429/19450. See in particular pp. 31–37.

2 **Glued laminated timber**, also called **Glulam**, is a type of structural timber product comprising a number of layers of dimensioned timber bonded together with durable, moisture-resistant structural adhesives, from Wikipedia. The BC Forest service pioneered the use of this material in bridge construction.

3 R.D. Bennett, "The Kingcombe Road," *British Columbia Lumberman*, (December, 1959), 66–67.

4 See Richard A. Rajala, "Forests and Fish: The 1972 Coast Logging Guidelines and British Columbia's First NDP Government," *BC Studies*, No. 159, (Autumn 2008), pp. 81–120. Rajala provides an in-depth study of many of the issues surrounding the policy and its regulation.

5 Richard Rajala. *Up-coast: Forest and Industry on British Columbia's North Coast 1870-2005,* Royal BC Museum, Victoria, 2006, p. 9.

6 Catherine Patrick to Dave Barrett. June 25, 1973, Forests, Lands and Natural Resource Operations, Narcosli Public Sustained Yield Unit, file 0272554.

7 "It's a Tragic and Frustrating Time for Crestbrook," *British Columbia Lumberman*, (July 1974), p. 22.

8 "Logging on an empty gut," *British Columbia Lumberman*, (July 1974, p.25.

9 Peter H. Pearse, *Timber Rights and Forest Policy in British Columbia*, Vol 1, Victoria, 1976, p. xi.

10 Jeremy Wilson, *Talk and Log: Wilderness Politics in British Columbia, 1965-96,* Vancouver; UBC Press, 1998, pp. 149–152.

11 Peter H. Pearse, *Timber Rights and Forest Policy in British Columbia*, Vol 1, Victoria, 1976. See pages 380–1.

12 http://www.for.gov.bc.ca/hts/pubs/tsr/tsa.htm

13 https://apps.gov.bc.ca/pub/geometadata/metadataDetail.do?recordUID=3729&recordSet=ISO19115

14 Examples of changes that have been implemented can be found on the BC Timber Sales website at http://www.for.gov.bc.ca/bcts/policy/

15 Joint Revenue Administration Review Team, Ministry of Forests, Treasury Board staff. "Report on the Revenue Administration Review of the Ministry of Forests," 1985.

16 Forest Service Policy Recreation Manual, Chapter 12, Wilderness Management, August 1991.

17 Fulton, Royal Commission Report, 1910.

18 Forests, Annual report 1923, E 10.

19 *Kalamalka Forestry Centre, Research and Seed Orchards*, Province of BC, Ministry of Forests, 1992, p. 6.

20 A.L. Orr-Ewing. Life History of the Deer Mouse, *Forestry Chronicle*. 1950,26(2): 115–126.

21 BC Ministry of Forests, *Forest and Range Resource Analysis* 1984. Victoria, British Columbia. 380p

22 R.A. Waldie, J. Selsing, W. Holland. "The Five Year Silviculture Plan for the Nelson region, 1980," unpublished plan Ministry of Forests, p16.

23 *Ibid.*

24 John Harkema and Mel Scott. "The Retention System: Maintaining Forest Ecosystem Diversity," Notes to the Field, Silvicultural System Program, Ministry of Forests, Vol 7, March 2002, p. 2.

Chapter 7

1 Fulton, Royal Commission Report, 1910

2 Ministry of Forests and Range Act [RSBC 1996] c. 300, section 4.

3 Gordon Hak. Capital and Labour in the British Columbia Forest Industry, 1934-74, Vancouver; UBC Press, 2007, p. 180.

4 Peter Pearse. Ready for Change, Crisis and Opportunity in the Coast Forest. A Report to the Minister of Forests on British Columbia's Coastal Forest Industry 2001, p. 35. http://www.for.gov.bc.ca/hfd/library/documents/phpreport/

5 Forest Resource Commission. The Future of Our Forests, Forest Resources Commission, Victoria, 1991, p. 86.

6 Marty Osberg and Brian Murphy. British Columbia Forest Practices Code, FAO corporate documents depository, http://www.fao.org/docrep/w3646e/w3646e0a.htm.

7 Forest Service Manual, Management Section, Instructions to Forest Officers.

8 Forest Service Manual, various dates.

9 Clark S. Binkley and Daowei Zhang. "Impact of Timber-Fee increases on British Columbia forest products companies: an economic and policy analysis", *Canadian Journal of Forest Research,* 28 617–625, 1998. John Clarke. "Logging Costs Skyrocketing in BC," *Logging and Sawmilling Journal,* 1997. http://www.forestnet.com/archives/May_97/soar.html

10 Jeremy Wilson. *Talk and Log: Wilderness Politics in British Columbia, 1965-96,* UBC Press, Vancouver, 1998, pp. 234–239.

11 The ministry definition of forest health: a forest condition that is naturally resilient to damage; characterized by biodiversity, it contains sustained habitat for timber, fish, wildlife, and humans, and meets present and future resource management.

12 Fulton, *Royal Commission Report, 1910,* p47.

13 Forests, Annual Report, 1915, I42.

14 Forests, Annual Report, 1934, U29.

15 T.C. Hargrave."All Fires Start Small," *British Columbia Lumberman,* (August 1951). 83–85, 91.

16 J.A. Turner, "Fire Weather Instruments and How to Use Them." *British Columbia Lumberman,* (May 1957), 25, 28–29.

17 The fire atlas is a set of maps which show the location of small fires and the boundaries of wildfires larger than 20 hectares. The cause of the fire is colour coded and the year included.

18 Robert H. Forbes. "Blitz Tactics Beat BC Blazes," *British Columbia Lumberman,* (September 1958), 12.

19 Keith Gibson. "Too Little! Too Late!" *British Columbia Lumberman,* (October 1958), 14–20.

20 A firestorm is a fire of such intensity that it creates its own internal windsystem.

21 Gary Filmon. Firestorm 2003, *Provincial Review,* Government of British Columbia.

22 Ministry of Forests, Backgrounder, "What is Results-Based Forest Management?," 2004, https://www.for.gov.bc.ca/code/backgrounders/whatis.pdf.

23 J.M. Swaine. "The Factors Determining the Distribution of North American Bark-Beetles," *The Canadian Entomologist,* LVII/II (November, 1925).

24 Ralph Hopping. "Forest Entomology," *Proceedings of the Entomological Society of British Columbia* Vancouver, October, 1923) p. 184.

25 Geo. R. Hopping and W.G. Mathers. "Observations on Outbreaks and Control of the Mountain Pine Beetle in the Lodgepole Pine Stands of Western Canada." *Forestry Chronicle* 21/2 1945 p. 98

26 Hector Allan Richmond. *Forever Green: The story of one of Canada's foremost foresters.* Lantzville: Oolichan Books, 1983, p. 93.

27 *The Tree Killer,* Province of BC, Ministry of Forests, 1981, p. 6.

28 C.B. Cottrell, L.S. Unger and R.L. Fiddick. *Timber Killed by Insects in British Columbia.* Environment Canada, Canadian Forest Service, Pacific Research Centre.

29 *A Review of Mountain Pine Beetle Problems in Canada,* Victoria: Environment Canada, Canadian Forestry Service, p. 8.

30 http://www.for.gov.bc.ca/ftp/hre/external/!publish/web/bcmpb/year8/BCMPB.v8.BeetleProjection.Update.pdf, p. 10.

31 Chief Forester's 2010 Annual Report on the Forest and Range Evaluation Program, February 2011, http://www.for.gov.bc.ca/ftp/hfp/external/!publish/frep/reports/FREP_Chief%20Forester's%20Report_Feb2011.pdf

32 Larry Pederson. "Allowable Annual Cuts in British Columbia: The Agony and the Ecstasy". UBC Faculty of Forestry, Jubilee Lecture, Vancouver, March 20, 2003, p. 23. Available at http://www.for.gov.bc.ca/hts/pubs/jubilee_ubc.pdf

33 Moving Toward a High Value, Globally Competitive, Sustainable Forest Industry. The Report of the Working Roundtable on Forestry, Province of British Columbia, March 2009. http://www.for.gov.bc.ca/mof/forestry_roundtable/Moving_Toward_a_Globally_Competitive_Forest_Industry.pdf

34 20110120-Roundtable Status Update, http://www.for.gov.bc.ca/mof/forestry_roundtable/

35 Ministry of Forests, Lands and Natural Resource Operations. Service Plan, revised 2011/12 – 2013/14 May 2011, http://www.bcbudget.gov.bc.ca/2011/sp/pdf/ministry/flnr.pdf, p. 7.

36 Forest Resources Commission. The Future of Our Forests, British Columbia, April 1991, pp. 25 and 27.

Epilogue

1 Announced via radio at 11 am October 25 2010. Field offices were caught by surprise. The functions of the service were divided with policy going to Ministry of Forest Lands and Mines and the balance to Ministry of Resource Operations.

2 Roger William, on his own behalf, on behalf of all other members of the Xeni Gwet'in First Nations Government and on behalf of all other members of the Tsilhqot'in Nation, Appellant, and Her Majesty The Queen in Right of the Province of British Columbia, Regional Manager of the Cariboo Forest Region and Attorney General of Canada. Tsilhqot'in Nation *v.* British Columbia, 2014 SCC 44 Date 20140626, Docket 3496.

3 http://www.princegeorgecitizen.com/news/legebokoff-trial/b-c-government-fraser-valley-first-nation-sign-new-revenue-sharing-deal-1.1203506.

Index

Page numbers in Italics refer to captions.